Navigating Agricultural Extension

About the Editors

Mr. Biswajit Mallick is currently pursuing Doctoral research in the Department of Agricultural Extension Education, Odisha University of Agriculture and Technology (Rank-3). He has qualified ICAR-NET-2023 and UGC-NET-2022. He has completed his Graduation from Centurion University of Technology and Management, Odisha. He joined Dr. Rajendra Prasad Central Agricultural University, Pusa, Bihar for Post Graduation through ICAR entrance exam as a Junior research fellow (JRF). He has research interest in areas like climate change, vulnerability, livelihood analysis. So far he has published six book chapters five popular articles and three research papers in NAAS rated National and International Journals.

Mr. Jyotishree Anshuman is currently pursuing Doctoral research in the Department of Agricultural Extension Education, Odisha University of Agriculture and Technology through ICAR entrance exam (State Rank-1), (ICAR Rank 18). He has qualified ICAR-NET-2023. He has also qualified UGC-NET and JRF 2023. He has completed his Graduation from Odisha University of Agriculture and Technology, Odisha. He joined Punjab Agricultural University, Ludhiana, Punjab for Post Graduation through ICAR entrance exam as a National Talent Scholar (NTS). So far he has published nine book chapters, three popular articles and two research papers in NAAS rated International Journals.

Navigating Agricultural Extension
A Comprehensive Guide

Editors

Biswajit Mallick

Jyotishree Anshuman

PhD Scholar

Department of Agricultural Extension Education

College of Agriculture

Odisha University of Agriculture and Technology hubaneswar

Bhubneshwar-751 003, Odisha

PMW

PMW
FU 75, Pitam Pura
New Delhi 110034
publishmywork23@gmail.com
Mob. 91-8447075807

ISBN: 978-81-96458-85-0

Preface

Agricultural extension is the lifeblood of sustainable farming, the bridge between knowledge and practice that empowers farmers to thrive in an ever-changing world. This book is a dedicated exploration of the critical field of agricultural extension, intended to serve as a valuable resource for extension professionals, educators, researchers, and anyone passionate about fostering agricultural progress. Agricultural extension profoundly impacts the lives of individuals and communities worldwide. It embodies the essence of collaboration, transferring knowledge and best practices from research centers and institutions to the fields and farms where they are needed most. Extension workers, often unsung heroes, connect farmers with the latest innovations, support systems, and tools that enable them to enhance productivity, adapt to challenges, and improve their livelihoods.In the pages that follow, we embark on a journey through the multifaceted world of agricultural extension, delving into its history, methodologies, and evolving role in the 21st century. We explore the dynamic relationship between extension professionals and the farming community, emphasizing the importance of effective communication, education and empowerment in achieving sustainable agricultural development.

This book is not just an informational resource but a call to action. As we confront the complex challenges of our time, from climate change and food security to market fluctuations and resource constraints, the role of agricultural extension becomes increasingly pivotal. We hope the insights and strategies presented herein will inspire readers to engage with agricultural extension in innovative and impactful ways, fostering resilient farming communities around the world. We extend our gratitude to the dedicated extension professionals who tirelessly work to bridge the gap between research and practice, helping farmers unlock their potential and overcome obstacles. Their commitment to knowledge dissemination and capacity building is a beacon of hope for global agriculture. As you delve into this book, we encourage you to approach its content with an open mind and a passion for advancing the field of agricultural extension. This book consists of 13 chapters contributed by MSc scholars and PhD scholars from different institutes throughout India. Whether you are a practitioner seeking new approaches, an educator preparing the next generation of extension experts, or simply someone interested in the transformative power of knowledge sharing, we invite you to explore the fascinating and ever-evolving world of agricultural extension. Thank you for joining us on this journey. May the knowledge and insights within these pages serve as seeds of inspiration, nurturing a brighter and more sustainable future for agriculture.

Happy reading!

Editors

Contents

1

Evolution of Agricultural Extension in Abroad and India

Debasish Pujari*, Deepa, Himanshu Yadav and Akhilesh Yadav

Research Scholar, Department of Agricultural Extension and Communication, Naini Agriculture Institute, Sam Higginbottom University of Agriculture, Technology and Sciences, Prayagraj (Allahabad) - 211007 Uttar Pradesh, India

Corresponding author's email: pujaridebasish580@gmail.com

Abstract

Extension is a key component in agricultural development, which is necessary for the introduction of new technology of farming to remote areas. Its first initiative was taken in between 1955-1970 and the thoughts were to increase the production in farming. The meaning of the term extension varied from country to country and language to language but its moral remains the same for all of them i.e. the term Voorlichting in Dutch, Perkembagan in America, Beratung in Germany and Forderum in Austrian are all the same but in their own way. It was first introduced in Cambridge University in 1873 and in very less time it spread in other parts of the world. From early 1950 the development of extension was always dependent on one or other factor such as either Govt. or some of the donor agency, where key facility like fertilizer, marketing and processing were missing, also Govt. policy were not helping the farmers as in later 1960s some of them got some new technology but still extension relied on poorly trained and overworked extension agents. In recent years, there has been a shift towards more participatory and decentralized extension approaches, which involve greater involvement of farmers in the extension process and greater collaboration between different stakeholders. This approach aims to better address the diverse needs and contexts of farmers and promote more sustainable and equitable agricultural development.

Keywords: *Agriculture Extension, Development, Farmers, Services, University.*

Introduction

Agricultural extension plays a pivotal role in the advancement of agriculture in any nation. It encompasses the dissemination of knowledge, guidance, and assistance to farmers, rural communities, and various stakeholders involved in the agricultural domain. The primary objective of agricultural extension is to empower farmers to enhance their productivity, income, and overall quality of life. The concept of agricultural extension has undergone significant transformations across various countries, including India. This article will delve into the historical progression of agricultural extension both internationally and within the Indian context.

Historical look of Extension

The initial phase of agricultural extension, spanning from 1955 to 1970, was primarily focused on establishing national agricultural extension systems. In the aftermath of World War II, there was a prevailing belief that the adoption of Western technology had the potential to significantly boost agricultural production. Central to this strategy was the education of farmers in the effective utilization of modern tools and technologies. Land grant university system started the model for transferring the technology (Thompson, 2002; USAID, 1998).

The interpretation of the term 'extension' has transformed over the course of time and carries distinct connotations in various nations. Across different historical periods and geographical regions, patterns of exchanging agricultural knowledge have persisted. In many instances, specific individuals such as religious leaders, traders, and elders have often assumed special advisory roles in facilitating this knowledge exchange. As per the findings of Jones and Garforth (1997), various forms of agricultural extension, which were somewhat institutionalized, can be traced back to ancient civilizations like Egypt, Mesopotamia, Greece, and Phoenicia. The term 'extension' has recent origin, stemming from academic circles. Its earliest documented usage can be traced back to Britain in the 1840s. During 1880s, it evolved into what was known as the 'extension movement,' where universities expanded their educational efforts beyond their physical campuses. In a similar vein, the term 'extension education' was employed in the United States since the early 1900s to signify that the scope of university teaching should not be confined to on-campus students but extended to individuals residing anywhere within the state. In this context, 'extension' is regarded as a variant of adult education, where the instructors are university staff members. Presently, most English-speaking nations have adopted the American term 'extension' to describe this concept. In contrast, other languages employ distinct words to depict similar phenomena.

Table 1: Meaning of Extension in different countries

Country/Origin	Word	Means
Arabic	Al-Ershad	Guidance
Austrians	Förderung (furthering)	Stimulating one to go in a desirable direction
Dutch	voorlichting	lighting the pathway ahead to help people
French	Vulgarisation	Extension
Germans	Beratung	Advisory work
India	Transfer of Technology	
Indonesia	penyuluhan	Giving illumination in dark
Korean	Rural guidance	
Malaysia	perkembangan	Development
Persian	Tarvij & Gostaresh	To promote & to extend
Spanish	Capacitacion	Training, Capacity building
Thai-Lao	Song-Suem	To promote
UK	Advisory work	Expert advice to select the way
USA	Education	To teach people to solve problems themselves

The concept of "Extension Education" was initially coined by Cambridge University in England in 1873. It referred to a distinct system focused on imparting knowledge to rural populations in their local environments, where they lived and worked. This idea quickly gained traction and spread to various regions in Britain, Europe, and North America. Extension work represents an extracurricular educational approach wherein both adults and young individuals engage in learning through practical, hands-on experiences. It embodies a collaborative alliance involving the government, land-grant institutions, and the community, aimed at delivering services and education tailored to address the specific requirements and aspirations of the people (Kelsey and Hearne, 1966). The term "Agricultural Extension" was officially adopted in 1914, coinciding with the enactment of the United States Federal 'Smith-Lever Act' of 1914. This legislation formalized a comprehensive nationwide partnership between federal, state, and county entities and entrusted the operational duties to the land grant colleges and universities. Initially, agricultural extension predominantly focused on enhancing agricultural practices through traditional teaching approaches. The expansion of agricultural extension into tropical Africa, the Caribbean, Asia, and Latin America was spurred by the involvement of the United States of America (USA) in bilateral AID programs following the conclusion of the Second World War.

Evolution of Agricultural Extension in Abroad

Agricultural extension has been an integral part of agriculture development in many countries. The history of agricultural extension can be traced back to the late 19th century when agricultural societies were formed in Europe and North America. These societies aimed to disseminate knowledge and information on agriculture to farmers through lectures, demonstrations, and publications (Swanson, 2010). In the early 20th century, agricultural extension services were established in many countries to provide farmers with technical assistance and advisory services. In the USA, the Cooperative Extension Service started after the 'Smith-Lever Act' in 1914. The Cooperative Extension Service provided farmers with education and technical assistance on various aspects of agriculture, including crop production, animal husbandry, and farm management (Swanson, 2010).

1. Evolution of Agriculture Extension in Europe

In Europe, agricultural extension services were established in many countries after World War II. These services were funded by governments and provided farmers with technical assistance and advisory services. In 1947, Agricultural Development and Advisory Service (ADAS) was established in United Kingdom, to provide farmers with advice on crop production, livestock management, and farm economics (Swanson, 2010). The evolution of agriculture extension services in Europe can be traced back to the 19th century when agricultural education and demonstration were the primary means of extension services. Agricultural schools and universities were established to provide education and training to farmers, and agricultural exhibitions and fairs were held to showcase new technologies and practices. The aim was to increase the knowledge and skills of farmers to improve their productivity and livelihoods. In the early 20th century, the focus shifted to advisory services, where experts provided advice and guidance to farmers on various aspects of agriculture, including crop management, animal husbandry, and soil conservation. This led to the establishment of agricultural extension services in many European countries, which were funded by the government or international organizations. During the mid-20th century, agricultural extension services expanded their scope to include rural development, marketing, and agribusiness. The focus was on improving the economic conditions of rural communities by promoting entrepreneurship and diversification. Extension services also played a critical role in the implementation of agricultural policies aimed at increasing food production and reducing poverty. In recent years, digital technologies have revolutionized agriculture extension services in Europe. Online platforms, mobile apps, and social media are being used to disseminate information, provide advisory services, and connect farmers with experts and other stakeholders. Precision agriculture and data analytics are also being used to optimize crop management and increase productivity. The use of digital technologies has made agriculture extension services more accessible and cost-effective, enabling farmers to access information and advice from anywhere at any time. It has also enabled extension services to reach a broader audience, including small-scale farmers who may not have access to traditional extension services (Davis and Winters, 2002).

Overall, the evolution of agriculture extension services in Europe has been driven by changing needs and priorities of farmers, advancements in technology, and government policies aimed at promoting sustainable agriculture and rural development. The future of agriculture extension services in Europe is likely to be shaped by new technologies, changing market demands, and the need to address environmental challenges such as climate change and biodiversity loss.

2. Evolution of Agricultural Extension in Australia

In Australia, the Department of Agriculture established a network of agricultural extension officers in the 1950s to provide farmers with technical assistance and advisory services. The extension officers worked closely with farmers to help them adopt new technologies and practices to improve their productivity and income (Swanson, 2010). Agriculture extension services in Australia have a long history dating back to the early 20th century. The first agricultural extension services were established in the 1910s, with the aim of improving the productivity and profitability of Australian farmers. In the early years, the focus was on providing farmers with advice and guidance on crop management, animal husbandry, and soil conservation. This was done through a network of government-funded agricultural advisers who worked closely with farmers to help them adopt new technologies and practices. During the mid-20th century, the focus of agriculture extension services in Australia shifted towards rural development and community building. This involved promoting entrepreneurship and diversification, as well as providing support for rural communities facing economic and social challenges. In the 1980s and 1990s, the Australian government introduced a range of policies aimed at promoting sustainable agriculture and environmental conservation. This led to a renewed focus on soil health, water management, and biodiversity conservation, with agriculture extension services playing a key role in promoting these practices. In recent years, digital technologies have transformed agriculture extension services in Australia. Online platforms, mobile apps, and social media are now widely used to disseminate information, provide advisory services, and connect farmers with experts and other stakeholders. Precision agriculture and data analytics are also being used to optimize crop management and increase productivity. This involves using sensors, drones, and other technologies to collect data on soil health, weather patterns, and crop yields, which can then be used to inform decision-making and improve agricultural practices. The use of digital technologies has made agriculture extension services more accessible and cost-effective, enabling farmers to access information and advice from anywhere at any time. It has also enabled extension services to reach a broader audience, including small-scale farmers who may not have access to traditional extension services. Overall, the evolution of agriculture extension services in Australia has been driven by changing needs and priorities of farmers, advancements in technology, and government policies aimed at promoting sustainable agriculture and rural development. The future of agriculture extension services in Australia is likely to be shaped by new technologies, changing market demands, and the need to address environmental challenges such as climate change and biodiversity loss.

3. Evolution of Agricultural Extension in Africa

In Africa, agricultural extension services were established in many countries in the 1960s and 1970s. These services were funded by governments and international organizations such as the Food and Agriculture Organization (FAO) and the World Bank. The aim of these services was to provide farmers with technical assistance and advisory services to improve their productivity and income (Swanson, 2010). Agriculture extension services in Africa have a long and complex history, shaped by colonialism, post-colonial development policies, and the challenges of poverty, food insecurity, and climate change. During the colonial period, agriculture extension services were primarily focused on promoting export-oriented cash crops, such as coffee, tea, and cocoa, while neglecting the needs of small-scale farmers who produced food for local consumption. This led to a legacy of inequality and dependence on external markets that continues to shape agricultural development in many African countries. After independence, many African governments established agriculture extension services as part of broader efforts to promote rural development and improve food security. However, these services often suffered from inadequate funding, poor infrastructure, and a lack of trained personnel, which limited their effectiveness in reaching small-scale farmers (Birner et al., 2007). In the 1980s and 1990s, the structural adjustment policies imposed by international financial institutions led to a reduction in government support for agriculture extension services, as well as the privatization of agricultural input markets (Shiferaw, et al., 2011). This led to a proliferation of private sector actors, such as agro-dealers and NGOs, who provided extension services to farmers. In recent years, there has been a renewed focus on strengthening public sector agriculture extension services in Africa, as part of efforts to promote sustainable agriculture and rural development (Davis, et al., 2010). This has involved investing in training and capacity building for extension workers, improving infrastructure and information systems, and promoting participatory approaches that involve farmers in decision-making. Digital technologies are also playing an increasingly important role in agriculture extension services in Africa. Mobile phones and other digital platforms are being used to disseminate information, provide advisory services, and connect farmers with markets and other stakeholders. Precision agriculture and data analytics are also being used to optimize crop management and increase productivity. However, the challenges facing agriculture extension services in Africa remain significant. These include limited funding and resources, poor infrastructure and connectivity, a lack of trained personnel, and the impacts of climate change and environmental degradation. Addressing these challenges will require sustained investment, political will, and a commitment to promoting equitable and sustainable agricultural development (Pretty et al., 2011)

4. Evolution of Agricultural Extension in India

Agricultural extension in India has a long history that can be traced back to the early 20th century. The first agricultural extension service in India was established in 1906 by the British government in Punjab. The aim of the service was to provide farmers with technical assistance and advisory services on crop production and animal husbandry (Singh, 2017). After independence, the government of India established the Directorate of Extension in 1953 to provide farmers with technical assistance and advisory services. The Directorate of Extension worked closely with state governments and other stakeholders to disseminate information on various aspects of agriculture, including crop production, animal husbandry, and farm management (Singh, 2017). In the 1960s, the government of India launched the Green Revolution program to increase food production and reduce poverty. The program involved the introduction of high-yielding varieties of wheat and rice, as well as the use of fertilizers and pesticides. Agricultural extension played a crucial role in the success of the Green Revolution by providing farmers with technical assistance and advisory services on the use of new technologies and practices (Singh, 2017). In the 1970s, the government of India launched the Intensive Agricultural District Program (IADP) to increase agricultural productivity in selected districts. The program involved the establishment of agricultural extension centers in each district to provide farmers with technical assistance and advisory services. The centers were staffed by extension workers who worked closely with farmers to help them adopt new technologies and practices (Singh, 2017). In the 1980s, the government of India launched the National Agricultural Technology Project (NATP) to promote the adoption of new technologies and practices in agriculture. The project involved the establishment of Krishi Vigyan Kendras (KVKs) in each district to provide farmers with technical assistance and advisory services. The KVKs were staffed by scientists and extension workers who worked closely with farmers to help them adopt new technologies and practices (Singh, 2017). In the 1990s, the government of India launched the Agricultural Technology Management Agency (ATMA) to decentralize agricultural extension services. The aim of ATMA was to involve farmers and other stakeholders in the planning and implementation of agricultural extension programs. ATMA was implemented in selected districts and involved the establishment of a district-level committee comprising representatives from various stakeholders, including farmers, NGOs, and government departments (Singh, 2017). In recent years, the government of India has launched several initiatives to modernize agricultural extension services. These initiatives include the use of information and communication technologies (ICTs) to disseminate information on agriculture to farmers. The government has also established a network of agricultural extension workers called Agriculture Extension Officers (AEOs) to provide farmers with technical assistance and advisory services (Singh, 2017).

Future of Agricultural Extension

The future of agricultural extension in abroad and India is likely to be influenced by various factors, including changes in technology, government policies, and the needs of farmers. In abroad, the use of ICTs is likely to play a crucial role in the dissemination of information on agriculture to farmers. The use of mobile phones, the internet, and social media platforms is likely to increase in the coming years, providing farmers with access to information on various aspects of agriculture. The establishment of public-private partnerships is also likely to increase, providing farmers with access to new technologies and practices. In India, the government is likely to continue its efforts to modernize agricultural extension services. The use of ICTs is likely to play a crucial role in the dissemination of information on agriculture to farmers. The government is also likely to focus on improving the quality of extension services by providing training and capacity building programs for extension workers. The involvement of farmers and other stakeholders in the planning and implementation of agricultural extension programs is also likely to increase, promoting a participatory approach to extension services. The future of agricultural extension in abroad and India is likely to be influenced by various factors, including changes in technology, government policies, and the needs of farmers. However, the core objective of agricultural extension will remain the same, which is to provide farmers with technical assistance and advisory services to improve their productivity, income, and overall well-being.

Conclusion

Agricultural extension has evolved over the years in different countries, including India. The evolution of agricultural extension has been influenced by various factors, including changes in technology, government policies, and the needs of farmers. In abroad, agricultural extension services were established in many countries in the early 20th century to provide farmers with technical assistance and advisory services. In India, the first agricultural extension service was established by the British government in Punjab in 1906. After independence, the government of India established the Directorate of Extension in 1953 to provide farmers with technical assistance and advisory services. Over the years, the government of India has launched several initiatives to modernize agricultural extension services, including the use of ICTs and the establishment of a network of AEOs. Agricultural extension will continue to play a crucial role in agriculture development in different countries, including India, by providing farmers with technical assistance and advisory services to improve their productivity and income.

References

Birner, R., Anderson, J., & Feder, G. (2007). Extension Services for Agricultural and Rural Development in Developing Countries: A Review of Innovative Practices. World Bank Policy Research Working Paper No. 4258.

Davis, K., & Winters, P. (2002). Agricultural Extension: A Step beyond the Cutting Edge. Food Policy, 27(4), 371-384.

Davis, K., Nkonya, E., Kato, E., Mekonnen, D., Odendo, M., Miiro, R., & Njuki, J. (2010). Impact of Farmer Field Schools on Agricultural Productivity and Poverty in East Africa. World Development, 38(10), 1384-1395.

Jones, B., & Garforth, C. (1997). Agricultural Extension: A Reference Manual. Rome: Food and Agriculture Organization of the United Nations.

Kelsey, M. and Hearne, R. (1966). Extension Education: A Reference Handbook. New York: McGraw-Hill.

Pretty, J., Toulmin, C., & Williams, S. (2011). Sustainable Intensification in African Agriculture. International Journal of Agricultural Sustainability, 9(1), 5-24.

Shiferaw, B., Hellin, J., & Muricho, G. (2011). Improving Market Access and Agricultural Productivity Growth in Africa: What Role for Farmer Organizations? Food Policy, 36(6), 713-725.

Singh, R. K. (2017). Agricultural extension in India: an overview. International Journal of Agriculture, Environment and Biotechnology, 10(4), 391-396.

Swanson, R. A., & Chermack, T. J. (2010). Strategy Development: Past, Present, and Future. Farnham: Gower Publishing Ltd.

Thompson, J. L. (2002). The Agricultural Extension System: A Look at Its Current Status and Future Directions. Journal of International Agricultural and Extension Education, 9(1), 5-14.

USAID. (1998). Agricultural Extension: The Next Step in Agricultural Development. Retrieved from https://pdf.usaid.gov/pdf_docs/PNABQ118.pdf

2

From Tradition to Transformation: The History of Agricultural Extension

Bijaylaxmi Behera[1*], Biswajit Mallick[2], BR Abha Ayushree[3] and Priya Singh[4]

[1,4]Research Scholar, Department of Extension Education and Communication Management,
Dr. Rajendra Prasad Central Agricultural University Pusa, Samastipur-848125, Bihar, India.
[2]PhD Scholar, Department of Agricultural Extension Education, College of Agriculture, Odisha University of Agriculture and Technology, Bhubaneswar-751003, Odisha, India.
[3]Assistant Professor, PG Department of Home Science, Rama Devi Women's University, Vidya Vihar, Bhubaneswar-751022.
**Corresponding author's email: bijaylaxmibehera671@gmail.com*

Abstract

Indian extension system has a rich history dating back to the early 20th century. The extension services in the country have played a crucial role in disseminating agricultural knowledge, providing technical assistance, and promoting modern farming practices among farmers. During the British colonial government in the 20th century, agricultural extension addressed the challenges faced by Indian agriculture, such as low productivity, pests, and diseases. Establishing agricultural departments, demonstration farms, and agricultural schools marked the beginning of formal extension services. After India gained independence in 1947, the government emphasized the development of cooperative societies to enhance the socio-economic conditions of farmers. Establishing agricultural universities and research institutes further strengthened the extension services through formal education and training. The 1960s witnessed the Green Revolution in India. The agricultural extension was pivotal in disseminating the knowledge and technology associated with the Green Revolution to farmers. The concept of "Transfer of Technology" was introduced in India. During the 1980s and 1990s, the focus shifted from a top-down approach to a more decentralized and participatory extension approach. The government encouraged the involvement of cultivators in the decision-making and implementation of extension programs. Non-governmental organizations (NGOs) and voluntary agencies also actively disseminate agricultural information and promote sustainable farming practices. Currently, efforts have been made to strengthen the linkage between research institutions, extension agencies, and farmers. Farmer-centric approaches, farmer field schools, and participatory rural appraisals have gained prominence. The focus has also shifted towards sustainable agriculture, organic farming, and climate-smart practices to address food security challenges, environmental sustainability, and climate change.

Keywords: *Agricultural Extension, Community Development, Developmental Programmes, First-line Extension, Five Year Plans.*

Introduction

Since about 10,000 years ago, men and women have been rearing livestock and producing crops. Information about agricultural practices can be found in documents dating back over 3,000 years from ancient Egypt, Mesopotamia, and China. Agricultural extension services have evolved and developed in many parts of the world to address agricultural systems' unique challenges and requirements. These initiatives aimed to provide farmers with scientific knowledge, technological advancements, and improved farming practices to enhance agricultural productivity.

Birth of Modern Agricultural Extension

In the mid-19th century, Ireland saw the creation of the first contemporary agricultural extension agency. The Irish potato crop was decimated with the spread of potato blight between 1845 and 1851, and a catastrophic famine occurred. The Royal Agriculture Improvement Society of Ireland (established in 1814) president received a letter from the British Viceroy assigned to Ireland in 1847 appointing "Itinerant Lectures" to travel to rural areas and instruct small farmers on how to grow nourishing root crops besides potatoes (Swanson, 2008). German government authorities were interested in the technique and set up their network of mobile teachers. By the ending of the 19th century, Italy, France, the Netherlands, and Denmark had adopted the concept. France's first nationwide, entirely state-run agriculture extension agency was founded in 1879. Agricultural organizations in the US organized agricultural fairs, exhibitions, and the sale of farm goods (The Berkshire Agricultural Society in 1811; The US Agril. Society in 1852). However, the Morrill Act (signed by President Lincoln) created Land Grant Colleges in 1862 to disseminate agricultural knowledge to farmers. Later, demonstration farms were established on college campuses. Before 1900, agricultural extension efforts had also begun in Japan. Following the Meiji Restoration in 1868, two agricultural institutions with faculty members from Europe were founded in the middle of the nineteenth century. Experimental work was done at these colleges and government farms. Agricultural fairs and displays began simultaneously, and forward-thinking Japanese farmers spoke to and demonstrated for the farmers. From 1881 onward, this resulted in the growth of numerous agricultural societies. The Japanese government also established a system of hiring seasoned farmers to serve as itinerant agricultural lecturers in 1885.

History of Extension Education in India

The rural economy was based primarily on agriculture until the British East India Company arrived. From 1875 until 1901, famines compelled the British government to establish certain commissions. 1871 the Departments of Revenue, Agriculture, and Commerce began operating. The famine committee report 1880 included a recommendation to form the agricultural secretariat, which was done in 1881. In most provinces, agricultural departments started working in 1882.

After the Famine Commission's report in 1901, the Provincial Department's scientific and developmental staff was strengthened. Greater financial support was given to the agricultural department, and full-time Directors of Agriculture were appointed in the major provinces under Lord Curzon's Viceroyalty (1898–1905). The provinces were divided into useful circles, and each circle had a deputy director of agriculture who oversaw operations related to experimental work and demonstration or extension activity. In 1906, the Imperial Agricultural Service was established. The Government of India Act, 1919 deemed agriculture development a state concern. The Royal Commission on Agriculture in India in 1928 recommended that demonstrations occur on cultivators' own fields rather than government demonstration farms and that the Director of Agriculture be appointed, whose sole responsibility would be to plan and coordinate propaganda efforts across the province.

Early Extension Effort before Independence

Significant attempts were undertaken to reconstruct rural communities before independence by several eminent social workers.

Scheme of Rural Reconstruction

Sir Daniel Hamilton experimented with model communities in Sunderban, Bengal region as early as 1903. A Central Cooperative Bank & Cooperative Marketing Society was established in 1924, and a Rural Reconstruction Institute was established in 1934, carrying on this work. Later, instruction for cottage enterprises was given (Adivireddy, 1997).

Sriniketan Experiment

Sri Rabindra Nath Tagore founded the Sriniketan Institute of Rural Reconstruction in Bengal in 1921. The program's focal point was a cluster of eight villages. The poet's inspiration was put to use in a variety of practical endeavors, including cooperative organization, agricultural demonstration, the provision of quality seed and effective manure, the development of cottage industries via the construction of a training center for handicrafts, the launch of the "Vratachari Movement" youth movement, and the establishment of a school named "Shiksha Satra," for teaching village boys and girls and the creation of study materials for illiterates.

Marthandam Experiment

Dr. Spencer Hatch headed the Marthandam Experiment, starting in Triandram in 1921. It was established under the Young Men's Christian Association (YMCA) guidance. The prime goal of the experiment was to foster the improvement of rural communities in terms of mental, physical, social, and economic development. Through YMCA centers, more than 100 villages were covered. It had treasured animals, equipment for the honey industry, and a demonstration field for enhancing crop and vegetable seeds. Some local businesses, weaving, raising chickens, and beekeeping, were founded on a cooperative and self-help foundation.

Gurgaon Experiment

In 1927, Mr. F.L. Brayne, the Collector of the Punjab district of Gurgaon, had the concept for rural development. Following this concept, a "village guide" was appointed in each village who served as a conduit for the villagers to receive guidance from professionals in many ministries. The program introduced improved seeds, tools, and cultivation techniques. Propaganda was spread through movies, music, and plays to raise farm productivity and health standards.

Sevagram Experiment

Mahatma Gandhi provided direction for the Sevagram experiment in 1933. But it took momentum in 1920 with the establishment of the All India Spinners Association. The primary goals were: (i) Service to the deprived; (ii) Decentralized production and equal distribution of wealth; and (iii) Self-sufficiency of Indian Villages. Efforts such as the construction of training centers for cottage enterprises, the promotion and practice of intergroup harmony, the outlawing of untouchability, and a focus on women's education made the experiment more viable.

Firka Development Scheme

The Firka development plan in Madras State was started in 1946 by Chief Minister Sri T. Prakasam in 34 Firkas all over the State. It was expanded to include another 50 Firkas on April 1st, 1950. It had a plan for rebuilding rural areas that drew inspiration from Mahatma Gandhi. The scheme's short-term goal was to create the institutional framework, essential utilities, and projects for communication, water supply, sanitation, and the creation of cooperatives and panchayats. The long-term goal was to establish khadi, cottage industries, animal husbandry, and agriculture to be self-sufficient in food, clothing, housing, etc.

Nilokheri Experiment

Another innovative effort that gained national notice and directly inspired the community development movement was started in 1948 by Sri S.K. Dey. About 7000 people forced to leave Pakistan during the partition were to be rehabilitated in the Nilokheri Township. Because it was based on the tenet that "he who would not work, neither shall he eat," the program was known as "Mazdoor Manzil." The goal was to achieve self-sufficiency regarding the necessities for the township and the nearby communities. The focal point of the activity was a vocational training facility.

Etawah Pilot Project

Etawah Pilot Project was started in 1948, with its headquarters in Mahewa hamlet, 11 miles outside Etawah, in the Uttar Pradesh district of Etawah. Mr. Albert Mayer of the United States designed the pilot project he launched in partnership with the Uttar Pradesh Government. The main goal was to determine the extent of social development, initiatives, confidence, and cooperation that can be created in the people. The pilot project in Etawah was deemed effective, and the methodology was adopted for community projects.

Grow More Food Campaign

The "Grow More Food Campaign" was a coordinated initiative to boost agricultural output in India in 1947. The GMF Enquiry Committee (1952) recommended (i) creating an extension organization for rural work and (ii) staffing each taluka with a block development officer (BDO), four number of technical staffs, and twelve village-level workers (VLWs) who work in an average of 120 villages.

Extension Efforts after Independence

Community Development

According to London School, "mass education" gave rise to the community development concept. The creation of community development as a government program began with the 1944 report on mass education in Africa by the Secretary of State's Advisory Committee on Education. The concept of "mass education" was dropped for various reasons. Later, the phrase "community development," which looked to have come from Cambridge, was used.

Definition of Community Development

This term was first proposed in 1948 by the Cambridge Summer Conference on African Administration. It defined it as a movement designed to promote better living for the entire community with active participation.

Community Development Programme in India

On October 2nd, 1952, the Community Development Program (CDP) was introduced with 55 Community Development Projects (CDP). With 300 villages and a population of roughly two lakhs, each project had an operational area of between 400 and 500 square miles. The project area was divided into three development blocks, each with between 60 and 70 thousand people and around 100 villages. A project officer and a number of extension officers with expertise in rural engineering, agriculture, animal husbandry, and other related fields oversaw the project. One Village Level Worker (VLW) was assigned to each group of five to ten villages, totaling around 60 VLWs for each project. The fundamental objective was the overall development of farming communities. While, its broad objectives were: (i) economic growth; (ii) social justice; and (iii) democratic development.

National Extension Service (NES.)

The National Extension Service (NES) program was introduced a year later, in 1953, to provide broader coverage at a lower cost and encourage greater public participation. Around 100 villages and 65 thousand people lived in each NES block. A Block Development Officer (B.D.O.) and many Extension Officers (EOs) were responsible for the NES block.

Panchayat Raj Institutions

The word "Panchayat" appears to come from the word **"Pancha"**, which means five. Panchayat, thus, denotes a "council of five." Traditionally, a panchayat is an assembly of elders who used to resolve conflicts and problems in a community or within a caste under local custom. Nobody dared to criticize their authority since it was rigid and customary. In its current iteration, the **Panchayat Raj** is a constitutional body and a decentralized unit of rural self-government that is democratically elected. It allows for the participation of the populace, especially the weaker groups within society, in local self-management.

Balvantray G. Mehta Committee Report

In order to investigate and provide a report on the CDP and NES, Balvantray G. Mehta committee was constituted in January 1957. The research team concluded that development was not possible except accountability and authority. Development of community took place when the local population was aware of its issues, accepts its obligations, exercises its authority through duly elected representatives, and continually monitors local government.

Balvantray G.Mehta committee recommended:

(i) Set up democratic organizations in village, block, and district levels through elected members;

(ii) Planning and development works assign to the elected bodies;

(iii) Provide available resources for smooth implementation of the developmental activities.

On behalf of suggestions recommended by the Team, on 2nd October 1959 Panchayati Raj was started in Rajasthan for the first time in the country, followed by Andhra Pradesh in November 1959 and Maharashtra on 1962. Subsequently, most of the states were adopted Panchayat Raj system.

In India, the Panchayat Raj Institutions typically follow a three-tier structure. Except for some States and Union Territories, the first level at the village level is Gram (Village) Panchayat. In Gujarat and Karnataka, it is known as a Taluka Panchayat; in Uttar Pradesh, a Janpad Panchayat; in Arunachal Pradesh, an Anchal Samiti; and Andhra Pradesh, a Mondal Parishad. The second level at the block level is typically referred to as Panchayat Samiti. Zilla Parishad, or District Panchayat, is the standard name for the third administrative level at the district level. Its sub-divisional name in Assam is Mohkuma Parishad.

Transfer of Technology: ICAR Projects and Models

According to Mutharamman (1994), technology transfer is the dissemination process of recent agricultural technologies developed in a research center or innovation system to the ultimate user (farmers) through an appropriate channel (extension system), which targets the specific group of farmers and motivate them to adopt and integrate the available technology into the present farming practices.

First-line Extension Projects

The ICAR launched Transfer of Technology Programmes at different times. These were:

1. National Demonstration Project- 1964
2. Operational Research Project- 1975
3. Krishi Vigyan Kendras- 1974
4. Lab to Land Programme- 1979

In 1971, ICAR made a department of Extension Education at its headquarters after understanding the extent and significance of an integrated relationship of research, education, and extension operations. This section was eventually enlarged and given the name Division of Agriculture Extension. ICAR was planned to implement this functional approach in agricultural universities, research institutes, and related organizations in the future.

1. All India Coordinated Project on National Demonstrations

In 1964, a nationwide demonstration program on essential food crops known as National Demonstrations (ND) was initiated. The idea was that even though the experts could back up their claims, the farmers might not take their counsel seriously. It was a cross-country undertaking with a standardized design and pattern.

2. Operational Research Project

After a decade of National Demonstrations, the Operational Research Project (ORP) gained reorganization in 1974. ORP aimed to disseminate the proven technology among farmers of whole villages or a cluster on a watershed basis (Prasad, Choudhary, & Nayar, 1987). The ORPs took into account two different types of issues: first, the widespread agricultural issues impacting the farming community that call for collective or communal action, such as plant protection and rodent control, and second, the overall resource development of the watershed area.

3. Krishi Vigyan Kendra (KVK.)

In 1974, ICAR introduced a new kind of experimental station and created history in the Indian agriculture system based on the suggestions of the Education Commission (1964–1965) and the Mohan Sinha Mehta Committee. With the foundation of the first Farm Science Center in Pondicherry, subsequently known as Krishi Vigyan Kendra, agricultural vocational training was provided to farmers. KVK is a non-mobile vocational training institute that imparts training to extension personnel, farmers, and the weaker section of society. It conducts various training programs, frontline demonstrations, and on-farm testing in the farmer's field.

4. Lab to Land Programme

In commemorating its Golden Jubilee, the ICAR introduced the Lab to Land Programme (LLP) in 1979. By transferring proven technology created by agricultural universities, research stations, etc., the program's overall goal was to boost the economic situation of small and marginal landholders, landless agricultural laborers, in particular, scheduled castes(SC) and scheduled tribes (ST).

Agricultural Development Programmes

During the post-independence period, Indian agriculture gained much momentum towards inclusive development. Five-year plans were started in the country after the formation of the planning commission on 15 March 1950 as an arm of the central govt. Initially, those plans focused on agriculture, industrial, economic, and rural development. There were several developmental programmes started during five-year plans. Some are mentioned below:

Developmental Programmes in FYPs- At a Glance		
Sl. No.	**Five Year Plans**	**Developmental Programmes**
1.	First Five-Year Plan (1951-1956)	Community Development Programme (1952), National Extension Service (1953)
2.	Second Five-Year Plan (1956-1961)	Intensive Agricultural District Programme (1960)

Developmental Programmes in FYPs- At a Glance		
Sl. No.	**Five Year Plans**	**Developmental Programmes**
3.	Third Five-Year Plan (1961-1966)	Intensive Agricultural Area Programme (1964), All India Coordinated Project on National Demonstration (1964), High Yielding Variety Programme (1966)
4.	Fourth Five-Year Plan (1969-1974)	Small Farmers Development Agency (1970-71), Marginal Farmers and Agricultural Labourers (1970-71), Drought Prone Area Programme (1970-71), Krishi Vigyan Kendra (1974)
5.	Fifth Five-Year Plan (1974-1979)	Training & Visit (1974)), Operational Research Project (1974-75), Training of Rural Youth for Self-Employment (1979), Lab to Land Programme (1979), Integrated Rural Development Programme (1978-79)
6.	Sixth Five-Year Plan (1980-1985)	National Rural Employment Programme (1980), Development of Women and Children in Rural Area (1982), National Agricultural Extension Project (1983), Rural Landless Employment Guarantee Programme (1983)
7.	Seventh Five-Year Plan (1985-1989)	Council for Advancement of Peoples Action and Rural Technology (1986). Technology Mission on Oilseeds (1986), Jawahar Rozgar Yojana(1989)
8.	Eight Five Year Plan (1992-1997)	Institute Village Linkage Programme (1995), Indira Awas Yojana (1996), Swarna Jayanti Sahari Rozgar Yojana (1997)
9.	Ninth Five-Year Plan (1997-2002)	Swarnajayanti Gram Swarozgar Yojana (1999), Jawahar Gram Samridhi Yojana (1999), Antyodaya Anna Yojana (2000), Sampoorna Gramin Rozgar Yojana (2001-02)
10.	Tenth Five-Year Plan (2002-2007)	Kisan Call Centre (2004), Providing Urban Amenities in Rural Area (2004), National Rural Employment Guarantee Act (2006)
11.	Eleventh Five-Year Plan (2007-2012)	National Food Security Mission (2007), Rashtriya Krishi Vikas Yojana (2007-08), Mahatma Gandhi National Rural employment Guarantee Act (2008), National Rural Livelihood Mission (2011)
12.	Twelfth Five-Year Plan (2012-2017)	PMKSY, PKVY, e-NAM, RADP, NMSA

Intensive Agricultural District Programme (IADP.)

IADP, also referred to as a "package program," was introduced to the nation starting in the 1960 kharif in seven districts: Pali (Rajasthan), Ludhiana (Punjab), Sahabad (Bihar), Thanjavur (Tamil Nadu), West Godavari (Andhra Pradesh), Aligarh (Uttar Pradesh), Raipur (Madhya Pradesh),

Intensive Agricultural Area Programme (IAAP.)

The government began to consider how it may spread the advantages of more advanced agricultural technology across a vast portion of the nation at a lower cost and with fewer staff members due to the IADP's outstanding successes in improved crop output. As a result, the IAAP was established in March 1964. This was a less intensive extension program than the IADP.

High Yielding Variety Programme (HYVP.)

The IADP and IAAP experiences led to the perception that high-yielding agricultural technology needed to be developed and improved immediately. The government acquired many high-yielding wheat varieties from the International Maize and Wheat Improvement Center (CIMMYT) in Mexico for trial, demonstrating their suitability for Indian conditions. Dr. Norman Borloug was responsible for this importation.

Drought Prone Areas Programme (DPAP.)

A special initiative for developing drought-prone areas, the DPAP, was created in 1971. Soil erosion, water stress, and environmental deterioration are present in locations susceptible to drought.

Training and Visit (T & V) Programme

By the middle of the 1970s, there were several issues with extension programs in developing nations. New approaches and tactics for extending farm technology were introduced to address this vulnerability in 1974.

National Agricultural Research Project (NARP.)

To enhance the regional research capacities of the State Agricultural Universities (SAUs), the Indian Council of Agricultural Research (ICAR) developed the National Agricultural Research Project (NARP).

Institute Village Linkage Programme (IVLP.)

1995 a pilot program for the Institute Village Linkage Programme (IVLP) was started. It was implemented in 42 centers of chosen State Agricultural Universities (SAUs) and Indian Council of Agricultural Research (ICAR) institutes. The idea is built on a participative mode, assuring links between scientists and farmers in a "bottom-up" approach.

National Agricultural Technology Project (NATP.)

The National Agricultural Technology Project (NATP) was started in 1998 by the ICAR. This project's Innovation in Technology Dissemination (ITD) component, which aims to create a system for technology transfer (TOT), has been proposed.

Integrated Rural Development Programme (IRDP.)

The IRDP was established in 1978–1979 to provide assets to the poor that generate income and self-employment opportunities.

Training of Rural Youth for Self-Employment (TRYSEM.)

A program supported by the Department of Rural Development of the Government of India introduced TRYSEM on August 15, 1979. TRYSEM was a supporting element of IRDP.

Rural Landless Employment Guarantee Programme (RLEGP.)

On August 15, 1983, the RLEGP officially launched for Women, SC and ST individuals, as well as landless laborers, received preference in the workplace.

Jawahar Rozgar Yojana (JRY)

On April 1st, 1989, the JRY employment initiative was launched to give the rural poor access to jobs. The merger of NREP and RLEGP formed this program. The Center and States split the program's expenses on 80:20 basis. The target demographic would be those who live in poverty. SC and ST applicants will be given preference for employment under the Yojana.

Indira Awaas Yojana (IAY)

The Indira Awaas Yojana is a federally funded program that offers homes to select groups of impoverished rural people who fall below the poverty line. Before coming to Jawahar Rozgar Yojana (JRY), it was a sub-scheme of the Rural Landless Employment Guarantee Programme (RLEGP). It has been put into effect as an independent plan since January 1996.

Swarnajayanti Gram Swarozgar Yojana (SGSY)

On April 1st, 1999, the Swarnajayanti Gram Swarozgar Yojana (S.G.S.Y.) was introduced in India by combining the previous I.R.D.P., T.R.Y.S.E.M., D.W.C.R.A., S.I.T.R.A., G.K.Y., and M.W.S. programs. In order to reduce poverty in rural areas, the all-encompassing SGSY program, which includes all the direction of self-employment, such as considering the poor people into SHGs, training, credits, infrastructure, and marketing strategies, was developed.

Development of Women and Children in Rural Areas (DWCRA.)

The flow of aid to women who are part of target group households has been extremely little, according to a midterm analysis of the execution of the Integrated Rural Development Programme (IRDP). As a part of the IRDP, the DWCRA program was launched in 1982–1983.

Mid-Day Meal Programme

The National Programme of Nutritional Support to Primary Education (NP-NSPE), a government-sponsored program, was introduced on August 15, 1995, to increase enrolment, retention, and attendance while improving children's nutritional status.

Mahatma Gandhi National Rural Employment Guarantee Act (MGNREGA.)

With the passage of the National Rural Employment Guarantee Act (NREGA) in 2005, every rural household with adults ready to perform unskilled manual labor is guaranteed 100 days of paid employment annually. As of February 2nd, 2006, the act is in effect in the notified districts. 200 districts were chosen for the first round of NREGA implementation. With effect from 1st April, 2008, NREGA has been extended to all districts in the country and renamed as Mahatma Gandhi National Rural Employment Guarantee Act (MGNREGA).

Conclusion

Agricultural extension in India has undergone significant transformations over the years, adapting to the varying needs of farmers and the agricultural sector. The efforts have aimed to enhance agricultural productivity, improve livelihoods, and promote sustainable and inclusive growth in the country's agriculture sector. In the pre-independence era, several social workers started various developmental programmes in India. After independence, five-year plans formulated by planning were the precursor of different programmes in the country. ICAR also contributed significantly towards education, research, and dissemination of new technologies to farming communities via. Its first-line extension programmes. Recently farmer-led, cooperative, and bottom-up extension approaches have prevailed in India.

References

Adivireddy, A. (1997). Extension Education: Bapatala, A.P.: Sri Lakshmi Press.

e-Krishi Shiksha. (2012). First line extension Programmes of ICAR.http://ecoursesonline.iasri.res.in/mod/page/view.php?id=65803

MANAGE. (2007). Introduction to Agricultural Extension Management. Published by *National Institute of Agricultural Extension Management,* Rajendranagar, Hyderabad. https://www.coursehero.com/file/18628290/aem101/

Mutharamman, P. (1994). Technology Transfer Programmes, Empowerment News Weekly, New Delhi, 19-25 December.

Narain, I. (1981). Democratic Decentralization: The Idea, the Image and the Reality in Panchayati Raj (ed) T.N. Chaturvedi and R.B. Jain IIPA, New Delhi.

Prasad, C., Choudhary, B. N. & Nayar, B. B. (1987). First-Line Transfer of Technology Projects. Indian Council of Agricultural Research, New Delhi.

Swanson, B. E. (2008). Global review of good agricultural extension and advisory service practices (Vol. 82). Rome: Food and Agriculture Organization of the United Nations. https://prollinova.kreisquadrat.com/wp-content/files/documents/resources/reading-materials/global20review20of20good20agri20fao202008.pdf

3

Critical Study on Farm Journalism

Shreya Kar[1*], Jyotishree Anshuman[2] and Sunidhi[3]

[1,3]Msc. Scholar, Department of Extension Education, College of Agriculture, Punjab Agricultural University, Ludhiana, Punjab, India
[2]Ph.D. Scholar, Department of Agricultural Extension Education, College of Agriculture, Odisha University of Agriculture and Technology, Bhubaneswar, Odisha, India
[*]Corresponding author's e-mail: shreyakar2015bd@gmail.com

Abstract

The chapter titled "Critical Study on Farm Journalism" delves into the crucial role that farm journalism plays in the context of agriculture and rural development. It begins by defining farm journalism as a specialized branch of journalism focused on reporting, analyzing, and disseminating information related to agriculture, farming, and rural affairs. The chapter provides historical insights, tracing the evolution of farm journalism and its significance in facilitating communication between agricultural stakeholders. Discussing the sources and types of farm journalism, the chapter explores the diverse mediums through which agricultural information is conveyed, including print, electronic, and digital media. It highlights the purpose of farm journalism, emphasizing its role in educating, informing, and advocating for the agricultural sector and he principles governing farm journalism. Furthermore, the chapter outlines the multifaceted functions of farm journalism, which encompass information dissemination, advocacy, market insights, sustainability promotion, rural development, and more. Lastly, the chapter elucidates the integral part played by farm journalism in agricultural and rural development, emphasizing its contribution to informed decision-making, policy advocacy, and the overall well-being of farming communities. In essence, this chapter underscores the indispensable importance of farm journalism as a catalyst for positive change in agriculture and rural areas.

Keywords: *Farm Journalism, Information, Media, News, Rural development*

Introduction

The world of agriculture stands at the crossroads of innovation and tradition, and at the heart of this dynamic sector lies the crucial role of farm journalism. In the backdrop of evolving technologies, shifting market dynamics, and the ever-pressing need for sustainability, farm journalism emerges as a vital conduit of information, education, and advocacy for farmers and rural communities. This chapter delves into the multifaceted realm of farm journalism, exploring its definition, scope, guiding principles, and its indispensable functions and roles in the sphere of rural development.

Definition

Agricultural/ Farm journalism can be characterized as a facet of societal engagement primarily focused on collecting, preparing, and disseminating news and opinions related to agriculture promptly to various audiences. This dissemination takes place through diverse communication channels such as print, electronic media (television and radio), and the internet.

A more specialized definition of agricultural journalism identifies it as a distinct branch of journalism that encompasses the practices of acquiring, composing, editing, and transmitting information through media outlets like newspapers, magazines, radio, television, advertising, and more. It also involves overseeing the managerial aspects linked to the production of such content (Jana, 2014). (Top of Form

History of Farm Journals in India

Agricultural journalism in India had its beginnings in the mid-19th century. During this period, three notable books were published, covering topics ranging from weather to European agriculture and agricultural implements. It was Sir Sayed Ahmad Khan, the founder of Aligarh Muslim University, who recommended to Aligarh collector Mr. J H Princip the establishment of a scientific society devoted to agricultural journalism. As a result, the first agricultural journalism society in India was founded at Aligarh Muslim University in 1859. This society played a crucial role in disseminating agricultural information sourced from books and periodicals to farmers, with publications being made available in languages such as Hindi, Bengali, and others.

In the late 19th century, the western part of India experienced severe drought conditions between 1887 and 1900. In response to this crisis, an American missionary published a work titled "Indian Famine" through the Imperial Agricultural Research Institute (IARI) located in Pusa, Bihar. Subsequently, formal agricultural publications began in 1906 at the Imperial Agricultural Research Institute (IARI), which included the launch of the "Agriculture Journal of India" and the "Memoirs of the Department of Agriculture in India." These publications focused on agriculture-related subjects and scientific research. In 1931, the publications and information divisions of the Indian Council of Agricultural Research (ICAR) introduced two new publications: the "Indian Journal of Agricultural Sciences," which delved into agricultural research, and the "Indian Journal of Veterinary Sciences," which focused on animal husbandry.The year 1948 marked the beginning of "Kheti," a publication dedicated to agriculture. In 1973, the ICAR expanded its publications with the introduction of two research journals and three semi-technical journals, including "Kheti" (monthly), "PhalPhool" (quarterly), "KrishiChayanika" (quarterly), "ICAR Reporter," and "ICAR News." (Sharma et al., 2018).

Additionally, state agricultural universities such as Ludhiana, Hisar, Jabalpur, Patna, Ranchi, Faizabad, Coimbatore, and Hyderabad, along with private publishers, rural development organizations, and non-governmental organizations (NGOs), initiated periodicals and farm journalism in 1973. However, despite these efforts, there remained a gap in fulfilling the information needs of farmers and generating employment opportunities.

Recognizing this gap, prominent newspapers like The Times of India, The Hindu, The Business Line, and The Business Standard started providing agricultural reports to farmers. In the same year, UNI began delivering daily agricultural news to farmers (Sharma, 1998).

It was in 1973 that technology made its foray into agricultural journalism in India. The technology of mass communication, including radio, television, telephones, computers, and the internet, witnessed significant innovations in the 20th century (David and Rene, 2006). These communication technologies played a pivotal role in providing farmers with employment opportunities and delivering critical information on climate, weather, optimal farming practices, and market conditions (Raju and Malagatti, 2017).

Sources of News

Farm journalism relies on a variety of sources to gather and report news related to agriculture, farming, and rural affairs. These sources provide journalists with the information they need to inform and educate their audience. Some common sources of news in farm journalism include:

Farmers and Agricultural Experts: Farmers themselves often serve as primary sources of information, sharing their experiences, challenges, and innovations. Agricultural experts, including agronomists, veterinarians, and extension agents, provide valuable insights and expertise.

Government Agencies: Government departments and agencies related to agriculture, such as the Department of Agriculture in various countries, provide data, policies, and reports on agricultural issues, including crop yields, market trends, and agricultural policies.

Research Institutions: Agricultural research organizations and universities conduct studies and experiments, producing valuable research findings that are often shared with the media. This information helps farmers stay updated on the latest advancements in farming practices.

Industry Associations: Agricultural industry associations and trade groups often release reports, market analyses, and updates on industry trends, which can be valuable sources of information for farm journalists.

Farm Publications: Magazines, newspapers, and online publications dedicated to agriculture regularly publish articles, features, and reports on various agricultural topics. Farm journalists may draw information and ideas from these sources.

Agricultural Conferences and Events: Journalists attend agricultural conferences, fairs, and events to gather firsthand information, conduct interviews, and report on emerging trends and technologies.

Online Databases and Websites: Agricultural websites, databases, and online platforms provide access to a wealth of agricultural information, including statistics, research papers, and news releases.

Government Reports and Data: Government agencies often publish reports, data, and statistics related to agriculture, which can be used as primary sources for farm journalism.

Farmers' Organizations and Cooperatives: Farmer associations and cooperatives are valuable sources of information, advocating for the interests of farmers and providing updates on agricultural policies and practices.

Environmental and Conservation Organizations: Organizations focused on environmental sustainability and conservation may provide information on topics like organic farming, sustainable practices, and the impact of agriculture on the environment.

Weather and Climate Data: Meteorological agencies supply weather and climate data that is essential for farmers, especially for crop planning and management.

Market Reports: Market data and reports from commodity exchanges and market analysts are vital for farmers and agribusinesses to make informed decisions about crop and livestock sales.

Farm journalists rely on a combination of these sources to provide accurate, timely, and relevant information to their readers, helping farmers and rural communities stay informed and make informed decisions.

Types of News

- Narratives preceding and following an event
- Accounts of personal experiences and achievements
- Recent advancements, including pest outbreaks, scientific breakthroughs, weather and crop updates, and advancements in ongoing projects
- Forecasts and projections, like extended agricultural and livestock reports, economic outlooks, and long-term weather predictions
- Thematic content that is linked to an event, circumstance, progression, or issue

Scope of Agricultural Journalism:

Agricultural journalism possesses an extensive purview, encompassing a wide array of subjects within the realms of agriculture, food production, and farming. Some of the principal domains addressed by agricultural journalism include:

Crop Cultivation: This entails the reporting of cutting-edge farming methods, optimal techniques, and inventive approaches in crop production. It also delves into subjects like irrigation methodologies, fertilizers, pest control, and soil quality management.

Animal Husbandry: Agricultural journalism also explores subjects connected to the care and management of livestock, including breeding, nutrition, and healthcare practices. It encompasses considerations related to animal welfare and the environmental consequences of livestock farming.

Food Production: This domain encompasses the examination of contemporary trends in food production, such as organic farming, sustainable agricultural practices, and food safety measures. It also scrutinizes matters concerning food security and the local agricultural repercussions of global food trade.

Rural Advancement: Agricultural journalism extends its reach to subjects pertaining to the development of rural areas, incorporating discussions on infrastructure enhancement, educational initiatives, healthcare services, and employment prospects. Additionally, it addresses concerns related to rural poverty and disparities.

Dwivedi (2023)

Principles

Principles guide to us to do work in right direction. Ray and Mondal (2005) mentioned some principles of farm journalism which are given below:

Use plain language: Explain technical terms briefly and simply, using everyday words with clear meanings.

Structure and order your arguments logically: Present your ideas in a clear sequence, making a clear distinction between the main points and less important details.

Keep main points concise: Stick to the key issues, focusing on achieving your stated goals without unnecessary verbosity.

Place crucial information at the beginning: Organize your content like an upside-down pyramid, with the most important information at the start.

Include lively and well-composed visuals: Use simple, striking pictures and photographs with good lighting and shading to effectively convey your message.

Craft an engaging piece: Make your presentation captivating, motivating, personal, and diverse enough to maintain the audience's interest.Top of Form

Purpose

- Share innovations and ideas emerging from fields, labs, scientific institutions, and research centers with the public.
- Discuss livestock marketing in various regions, aside from agricultural technologies.
- Publish information about droughts, floods, and other weather-related facts, along with the responsibilities of different population groups.
- Explain government agricultural policies periodically.
- Share articles related to crop and livestock processing.
- Write about village health and cottage industries.
- Explore the social connections in rural life.
- Enhance people's understanding of agricultural extension issues within the rural economy.
- Develop writing skills to communicate various aspects of agricultural development in simple language.
- Plan and timely report on agricultural news, research results, and the challenges and opportunities of modern agricultural technologies to farmers.

Singh (2013)

Functions of Journalism

News Function: The primary role of the press is to provide information. When sifting through the abundance of public events, ideas, and situations, journalists must decide which ones will capture the public's interest. In addition to presenting the facts, complex situations often require interpretation and explanation.

The Opinion Function: In the modern world, people often find themselves facing confusion: What product should they buy? Which decision should they make?

Whom should they vote for? They need a communication medium that can help them comprehend the pros and cons of situations through logical arguments. Thus, today's press serves as both a daily educator and a platform for expressing viewpoints. This is why editorials play a pivotal role in shaping public opinion.

The Entertainment Function: Providing entertainment to the public is both a function and a business for the press. Due to the enormity of this task, newspapers often rely on syndicated content. Entertainment can be found in various forms, such as human-interest stories and feature articles. Sometimes, a newspaper's choice is influenced by the public's interest in specific features, especially comics.

Training and Coordination Function: This function involves training individuals in agricultural information communication principles and maintaining close connections with newspapers, magazines, All India Radio (AIR), and Doordarshan for the dissemination of agricultural information.

Anonymous (2015)

Role of Journalism in Agriculture / Rural Development

- In the realm of agricultural and farm journalism, it's recognized that timely and well-presented technical information is essential for farmers to adopt innovations promptly. For instance, during the COVID-19 pandemic, farm journalists played a pivotal role in sharing crucial details about safety measures, government policies, and market conditions with farmers.
- Agricultural journalism draws on various major disciplines, including Extension Education, Mass Communication, Anthropology, Rural Sociology, and others.
- Publications centered on knowledge must not only provide information but also influence the reading habits and perspectives of their audience. For instance, programs like VV and the Farm and Home unit of AIR educate by integrating Knowledge, Skills, and Attitudes to address real-life problems and find appropriate solutions.
- Farm journalism aims not merely to convey information but also to guide the development of skills and competencies among specific clientele groups. Farm journalists can empower farmers by imparting knowledge about contemporary agricultural practices, crop management, and technological advancements.
- Through investigative reporting, farm journalists can illuminate agricultural policies in need of reform and advocate for changes that benefit farmers.
- Methods and challenges related to reaching and influencing farmers and others are receiving increased attention, with farm journalism playing an effective role in this context.

- Farm journalists offer valuable insights into market dynamics and price trends, enabling farmers to make informed decisions regarding the timing and location of their produce sales.
- Farm journalism can promote awareness of sustainable farming practices, the conservation of natural resources, and the impact of climate change on agriculture.
- Farm journalism contributes to rural development by showcasing success stories, community initiatives, and government programs aimed at enhancing rural livelihoods.
- Farm journalists educate farmers about the latest advancements in crop protection, disease management, and livestock health, thereby enhancing agricultural productivity.
- Through farm journalism, urban readers gain insight into rural issues, bridging the gap between urban and rural communities.
- Farm journalism provides a platform for farmers to share their experiences, challenges, and innovative solutions, amplifying their voices in public discussions.
- Journalists often collaborate with agricultural researchers to convey innovations and breakthroughs to farmers effectively.
- The demand for farm journalists is on the rise due to the diverse array of organizations and institutes, including farming publishers, farm associations, input agencies, farm cooperatives, radio/TV stations, advertising agencies, and dairy organizations, heavily relying on their services.Top of Form

Bhaskran et al. (2008), Ram Datt (2010), Jana & Mitra (2005), Pashudhan (2020)

Conclusion

The chapter on the "Importance of Farm Journalism" underscores the pivotal role this specialized branch of journalism plays in the realm of agriculture and rural development. Farm journalism, defined as the dissemination of agricultural information through various media, emerges as a critical conduit of knowledge. Tracing its historical evolution, we see how farm journalism has evolved to become an indispensable resource for farmers, policymakers, and the broader public. It sources information from a multitude of channels and employs various types of media to reach its diverse audience.The chapter emphasizes that farm journalism's purpose extends beyond mere information sharing; it is a dynamic force driving agricultural progress. Guided by principles of accuracy and impartiality, it functions as an educator, advocate, and promoter of sustainability. Through its multifaceted roles, farm journalism bridges gaps, empowers farmers, influences policy changes, and contributes significantly to rural development. In essence; this chapter underscores the undeniable importance of farm journalism in fostering agricultural growth, empowering rural communities, and promoting a sustainable future for agriculture.

References

Anonymous (2015). Journalism. Retrieved from http://ecoursesonline.iasri.res.in /Courses/ Extension%20Methodologies%20for%20Transfer%20of%2 0ag.%20Tech/Data%20 Files/lect14.html.

Bhaskran et al., (2008). Farm Journalism and Media Management. Agro tech publishing company.

David, L. D. and Rene, P. M. (2006). What Are Agriculture Industry Professionals Trying to Tell Us? Implications for University-Level Agricultural Communications Curricula. Journal of Applied Communication, 90(3), 1-18.

Dwivedi, S. (2023). Agricultural Journalism: Exploring Vital Role of Informing and Educating on Farming, Food Production & Industry Issues. Krishi Jagran. Retrieved from https://krishijagran.com/blog/agricultural-journalism-exploring-vital-role-of-informing-and-educating-on-farming-food-production-industry-issues/

Jana, B. L. (2014). Agricultural Journalism. Agrotech Pub. Academy.

Jana B.L & K.P. Mitra (2005). Farm Journalism, Agro tech publishing academy.

Pashudhanpraharee.(2020). Importance of farm journalism in livestock development. Retrieved from https://www.pashudhanpraharee.com/importance-of-farm-journalism-in-livestock-development/

Raju, K. &Malagatti, S. M. (2017). Agricultural journalism is an effective tool for farmers development in Dharwad district of Karnataka state. Rec. Res. Soc. Sci. Hum, 4(3), 66-70.

Ram, Datt. (2010). Handbook of Farm Journalism, Pulari Publishers.

Ray, G. L. &Mondal, S. (2005). Journalism Including Communication Farm & Rural Journalism Public Relations.Kalyani Publishers.

Sharma, A., Sutradhar, M., Monlai, S., &Kumari, N. (2018). Agriculture journalism brings employment in India. International Journal of Current Microbiology and Applied Sciences, 7(09), 2314-2319.

Sharma, R. D. (1998). Growth of Agricultural Journalism and Agricultural Research Information Centre at ICAR.DESIDOC Bulletin of Information Technology, 18(2), 21-30.

Singh, C. B. Thamizhan. (2013). Farm Journalism. Term Paper. Retrieved from https://www.termpaperwarehouse.com/essay-on/Farm-Journalism/507975

4

Digital Agriculture: Exploring the Role of Information and Communication Technology for Sustainable Development

Pallavi Shaktawat* and Sharmistha Swaymprava

Research Scholar, CPGS-AS, Central Agricultural University
Umiam, 793103, Meghalaya
**Corresponding author's email: Pallavishaktawat1@gmail.com*

Abstract

ICT (Information and Communication Technology) has emerged as a pivotal driver in agricultural development. ICT provides farmers with up-to-the-minute information on weather, soil conditions, and crop health, enabling careful resource management to allow farmers to make more informed decisions, better utilize resources, and increase agricultural yields. This technology promotes sustainable practices, market access, and knowledge sharing, which helps rural communities prosper economically. For widespread implementation, however, problems with connectivity and computer literacy must be resolved. This chapter emphasizes the need for continuing ICT integration in order to secure rural development, food security, and resilience in an increasingly dynamic farming landscape. In the future, the convergence of 5G, AI, and blockchain is expected to transform Indian agriculture and pave the way for a data-driven, effective, and resilient society.

Keywords: *Blockchain, ICT, Indian Agriculture, Modern Technology.*

Introduction

India's vast and diverse agricultural landscape has one of the greatest agricultural sectors. With over 60% of its people working in agriculture, India's economy and food security heavily depend on agriculture. In recent years, the world has witnessed many difficulties, including information access restrictions, dispersed property ownership, and uncertainty brought on by climate change. Information and Communication Technology (ICT) has emerged as a potent tool for empowering farmers, boosting production, promoting sustainable practices in response to these difficulties, and bringing game-changing transformations in Indian agriculture. India ranks second among Asian countries in the advanced user category. On the one hand, profit motivation, corporate expansion, community services, and rural welfare have been the goals of ICT-based agricultural models in India (Spielman *et al.,* 2021).

This chapter explores how ICT can revolutionize agriculture, leading to sustainable development, increased productivity, and improved livelihoods for farmers worldwide (Swaminathan and Swaminathan, 2018). ICT includes a broad range of technologies used to develop cutting-edge agricultural solutions, such as sensor networks, data analytics, mobile applications, and satellite images. This chapter will explore the various strategies that have influenced the agricultural environment in India.

1. ICT in Agriculture: An Overview

India's foray into the realm of ICT can be traced back to the early 1980s when the government took proactive measures to foster the growth of indigenous computer systems. A significant milestone was set in motion during the 1990s with the establishment of the Software Technology Parks of India (STPI). This pivotal development catalyzed the expansion of India's IT industry. It was during this era that Indian powerhouses with IT companies such as Tata Consultancy Services (TCS), Infosys, and Wipro emerged on the global stage, laying a solid foundation for India's subsequent dominance in the field of information technology.

Information and communication technology (ICT) plays a significant part in agricultural growth development by utilizing various gadgets to achieve economic sustainability and self-sufficiency (Sennuga *et al.,* 2020). Farmers across India can now access mobile apps and web platforms that provide real-time weather forecasts, market prices, and expert advice. This information empowers them to make informed decisions regarding crop selection, planting times, and selling strategies, ultimately improving crop yields and income. It also facilitated the implementation of precision agriculture techniques, such as remote sensing and GPS-guided machinery, allowing farmers to optimize resource usage, reduce wastage, and minimize environmental impact. E-agriculture disseminates important information by incorporating the User Interface and Knowledge Management via a Decision Support System (DSS), Management Information System (MIS), and Expert System (ES) (Saidu *et al.,* 2017; Behera *et al.,* 2015).

1.1 The Role of ICT in Agriculture

The Role of Information and Communication Technology (ICT) in agriculture is becoming more important because it helps to modernize, streamline, and make the agricultural industry more sustainable. ICT includes many technologies, such as mobile apps, sensors, drones, and data analysis. These tools will help farmers improve their farming practices by providing real-time data, actionable insights, and innovative solutions (More and Aslekar, 2020). Here are some key aspects of ICT's role in agriculture:

1.1.1 Access to Information: ICT tools provide farmers timely access to vital information. Farmers can easily access weather forecasts, market prices, pest and disease management strategies, and best agricultural practices through mobile apps, websites, and SMS services. This information empowers farmers to make informed decisions and adapt to changing conditions.

1.1.2 Precision Agriculture: Using ICT, precision farming techniques are made possible. This involves collecting data from satellites, drones, and sensors to optimize the use of resources. Farmers can use technology to better control irrigation, fertilization, and pest control. This helps them save money, protect the environment, and increase their crop yields.

1.1.3 Financial inclusion: Thanks to mobile banking and electronic payment systems, farmers can access safe and practical financial services. These cover having access to financing, insurance, and subsidies—all of which are essential for investing in contemporary agricultural methods.

1.1.4 Data Analytics: With the help of ICT, we can gather and analyze a lot of agricultural data. Machine learning and AI can help us understand the health of crops the quality of soil, and predict crop yields. This information can assist in making decisions and managing risks.

1.1.5 Climate Resilience: ICT tools help farmers adapt to changing climate conditions in agriculture by providing early warning systems for extreme weather events.

1.1.6 Supply Chain Management: ICT systems can track the movement of agricultural products along the whole supply chain. This transparency reduces waste, stops food from spoiling, and keeps product quality high.

1.1.7 Capacity Building: Online resources and training help farmers learn new skills and stay updated on agricultural innovations. Virtual training programs and webinars link farmers with experts and peers to share knowledge.

1.1.8 Policy and Governance: Governments use ICT to implement agricultural policies, distribute subsidies, and monitor the sector's performance. It enhances transparency and reduces corruption in agricultural governance.

1.2 Benefits of ICT in Agriculture

1.2.1 Precision Farming: ICT makes precision agriculture possible by supplying precise information on crop health, weather patterns, and soil conditions. Farmers are better equipped to maximize resource use, cut waste, and boost yields.

1.2.2 Remote Monitoring: With the use of IoT (Internet of Things) sensors and cameras, farmers can remotely monitor their crops and livestock. This makes it possible to identify illnesses, pests, and other problems early, which enables speedier response times and healthier plant and animal life.

1.2.3 Market Access: Online marketplaces and e-commerce platforms help farmers connect directly with buyers. Farmers can sell their products to customers across the country and around the world. This allows them to bypass middlemen and get higher prices for what they grow.

1.2.4 Knowledge Sharing: Farmers can exchange information and best practices via social media, smartphone apps, and online forums, which encourages creativity and lifelong learning.

1.2.5 Sustainable Practices: Through recommendations based on data, ICT aids sustainable agriculture by promoting techniques like organic farming, crop rotation, and reduced pesticide use.

1.2.6 Economic Growth: The use of ICT in agriculture increases farmer income and fosters economic development in rural areas.

2.1 Sensor Networks and Precision Agriculture

Sensor networks and precision agriculture have significantly transformed Indian agriculture, offering precise and data-driven approaches to farming. These technologies empower Indian farmers to make informed decisions, optimize resource usage, and enhance crop yields. Here, we'll delve into specific examples of how sensor networks and precision agriculture are making a difference in India:

2.1.1. Soil Health Monitoring Sensor Networks: They are used to analyze soil parameters such as moisture content levels and pH of soil. Farmers receive real-time data through mobile apps or web platforms to help them determine optimum nutrient levels and water required for irrigation. For example, farmers in Punjab use soil sensors to measure water levels, enabling irrigation and reducing water wastage.

2.1.2. Weather Stations: Weather sensors deliver precise weather forecasts and individualized weather forecasts. Indian farmers use this information to schedule planting and harvesting operations keep an eye on temperature extremes and prepare for rain Weather sensors are used nationwide by organisations like the Indian Meteorological Department (IMD) to supply farmers with timely weather information.

2.1.3. Precision irrigation: Irrigation systems are controlled by sensors to guarantee that crops receive the appropriate amount of water at the appropriate time. For instance, farmers in Telangana use IoT-based irrigation systems to optimize water use in rice fields. This helps to conserve water resources and boost crop yields.

2.1.4. Market Intelligence: Mobile apps and online platforms provide market intelligence by gathering data on commodity prices and demand. Farmers may make educated choices about when and where to sell their produce. Agri-tech Startups such as AgriBazaar, AgmarkNETand Ninjacart operate in multiple states, connecting farmers with buyers directly.

2.1.5. Pest and Disease Management: Sensor networks can identify the presence of pests and illnesses in fields using traps and sensors. Farmers get quickly notified when insect populations reach a specific threshold, along with suggestions for interacting with experts to prevent further damage to crop. In Andhra Pradesh, for example, sensors are being used to track the progress of the autumn armyworm in maize fields.

2.1.6. Government Initiatives: The Indian government has created efforts to encourage precision agriculture, such as the National Mission on Agriculture Extension and Technology (NMAET). Subsidies for precision farming equipment are included to encourage widespread adoption.

2.1.7. Livestock Monitoring: Wearable devices with sensors are used to monitor the health and behavior of animals. These devices monitor parameters like body temperature and movement, allowing for early diagnosis. Indian dairies, especially in Gujarat and Haryana, have adopted these sensors to improve animal health and milk production.

2.2 Drones in Agriculture

Drones provide creative answers to farmers' difficulties in India. Drones outfitted with high-resolution cameras and multispectral sensors are widely utilized in India to monitor crops, analyze soil health, and improve agricultural practices. For example, farmers use drones in places like Maharashtra, Meghalaya, and Karnataka to monitor crops like cotton and sugarcane and enable focused interventions, allowing farmers to spot areas of stress, pest infestations, or nutrient deficits. This information is then analyzed using advanced analytics and artificial intelligence to provide farmers with actionable insights to help them make informed decisions about the nutrient application and irrigation schedule, thereby increasing crop yields and optimizing resource utilization.

Drones also help to address one of the most pressing issues in Indian agriculture: small and fragmented land holdings and tackling the issue of farm size in India. Drones provide digitized land records with improved mapping and surveying capabilities, assisting farmers to establish property boundaries and access government schemes and subsidies. This is critical for ensuring equal resource distribution and encouraging rural development. The Indian government has recognized the potential use of agricultural drones and has implemented rules and incentives to encourage their use. Drones are used for crop insurance assessments in programmes such as the Pradhan Mantri Fasal Bima Yojana (PMFBY). Furthermore, the "Digital Sky" government initiative aims to regulate and integrate drones into the national airspace system, easing their use in various industries, including agriculture. GoI, (2022) has simplified restrictions to simplify drone integration into agriculture practices by the Directorate General of Civil Aviation (DGCA).

In addition to crop monitoring, Drone is utilised for developing detailed maps of the terrain and soil conditions for making precise sowing and fertilization possible. This guarantees resources such as seeds, fertilizers, and water are dispersed efficiently, decreasing waste and environmental impact. Drones can reach difficult-to-reach places and deliver precise treatment, making them an excellent alternative for Indian agriculture. Furthermore, in areas such as Punjab, drones aid in the proper dispersion of fertilizers and pesticides and are also aid in irrigation control, allowing farmers to maximize water utilisation. Drones equipped with thermal cameras detect differences in soil moisture and assist efficient irrigation practices in water-scarce regions such as Rajasthan, where agriculture faces severe challenges due to restricted water resources.

Challenges and Future Directions

3.1 Challenges in Implementing ICT in Agriculture

Implementing Information and Communication Technology (ICT) in Indian agriculture holds tremendous potential, but it also comes with several challenges that must be addressed for widespread adoption and success. Here, we'll elaborate on four key challenges:

3.1.1 Connectivity Issues

Connectivity remains a critical barrier to efficient ICT deployment in Indian agriculture, particularly in remote rural areas. Many rural towns continue to lack consistent internet connectivity and mobile network coverage. Farmers confront challenges obtaining online information, using mobile apps, and utilising IoT devices without a stable internet infrastructure. This problem is especially serious in regions such as NEH Regions, Jharkhand and Chhattisgarh, where mountainous terrain and extensive forests can obstruct connectivity.

3.1.2 High Initial Costs

Many small-scale farmers in India are put off by the hefty initial expenses of implementing ICT technologies. Investing in hardware such as sensors, drones, precision agricultural equipment, and software and training can be costly. This problem is particularly acute in areas such as Bihar and Uttar Pradesh, where a huge majority of farmers lack access to finance and financial resources.

3.1.3 Data Privacy and Security

In the digital age, data privacy and security issues are paramount. Farmers are concerned about the misuse of agricultural data, which frequently contains sensitive information about their property, crops, and harvests. Inadequate data protection methods pose data security hazards and lack awareness about data security practices. It is critical to ensure the safe and ethical use of farmers' data, especially in places like Punjab and Haryana, where data-driven agriculture is more prominent.

3.1.4 Digital Literacy among Farmers

Another key concern is a lack of digital literacy among farmers. Many farmers, particularly in regions such as Odisha and West Bengal, may be unfamiliar with cell phones, applications, or internet platforms. They may struggle to make the most of available ICT tools and resources if they lack basic digital skills and training. Bridging the digital literacy gap is critical for farmers to be able to use technology successfully for their benefit.

3.2 Future Directions

3.2.1 5G and Edge Computing

The combination of 5G technology and edge computing holds enormous promise for Indian agriculture's future. Farmers can access and transfer data in real-time thanks to 5G networks' ultra-fast and low-latency connectivity, enabling the deployment of advanced IoT sensors and devices in remote rural locations. Edge computing minimises latency and improves the responsiveness of agricultural applications by processing data closer to the source. Precision agriculture innovations, such as autonomous tractors, drones, and smart irrigation systems, could be widely used in India. These technologies can give farmers real-time information about soil conditions, crop health, and weather forecasts, allowing for better informed decision-making and resource management.

3.2.2 Artificial Intelligence and Machine Learning

AI and machine learning (ML) have the potential to play vital roles in the modernization of Indian agriculture. AI systems can analyse massive datasets from sensors, satellites, and drones to deliver useful knowledge. Machine learning algorithms can forecast crop yields, detect disease outbreaks, and recommend optimal planting and harvesting dates. AI-powered chatbots and smartphone apps can provide farmers with personalised assistance, even those with little literacy. AI and machine learning applications are predicted to revolutionise crop management in India, allowing farmers to enhance yields, lower costs, and limit risks while contributing to food security.

3.2.3 Blockchain for Supply Chain Transparency

Blockchain technology has the power to address significant issues in the agricultural supply chain while also assuring transparency and trust among stakeholders. Blockchain can trace agricultural products from farm to fork in India, where the supply chain can be fragmented and prone to inefficiencies and fraud. This transparency benefits farmers as well as customers by decreasing spoilage, eliminating counterfeiting, and ensuring fair pricing. Blockchain, for example, can verify the legitimacy of organic products, allowing producers to command higher prices. It can also improve farmers' revenue by streamlining payments and reducing dependency on intermediaries.

3.2.4 Climate-Resilient ICT Solutions

Climate variability and extreme weather events are rising in India, causing considerable difficulties to agriculture. Adaptation requires climate-resilient ICT solutions. These include sophisticated weather forecasting models, drought and flood early warning systems, and novel irrigation tactics. Remote sensing technologies combined with artificial intelligence (AI) can monitor soil moisture and assess crop stress caused by changing weather patterns. India can secure the sustainability of its agriculture, improve disaster preparedness, and protect food production in the face of climate change by deploying climate-resilient ICT solutions.

ICT Initiatives in Indian Agriculture

1. Digital Farming Platforms

Digital farming platforms have gained prominence in India, connecting farmers with critical information on weather forecasts, crop advisories, market prices, and best agricultural practices. Digital farming platforms act as comprehensive ecosystems that bring together various ICT tools and services for farmers:

- Government Initiatives: The Indian government has launched initiatives like the National Agriculture Market (e-NAM), which provides a digital marketplace for farmers to sell their produce nationwide. Table 1 provides an overview of e-NAM's impact on Indian agriculture.

Table 1: Impact of e-NAM on Indian Agriculture

Sl. No.	Metric	2015-16	2021-22	Change (%)
1.	Number of Mandis Integrated	250	1,000+	+300
2.	Volume of Trade (in crore ₹)	5,215	45,157	+766
3.	Farmer Beneficiaries	30,000+	1,000,000+	+3233

(Source: Ministry of Agriculture and Farmers Welfare, Government of India, 2020-21)

Private Sector Investments: Companies such as as ITC e-Choupal have developed digital platforms that provide farmers with access to information, inputs, and market connections, thereby promoting rural development.

- Data Analytics and Decision Support: These platforms frequently integrate data analytics and decision support systems, allowing farmers to make data-driven planting, harvesting, and market timing decisions.

2. Precision Agriculture

Precision agriculture employs technology to optimize resource utilization and increase crop yield. It involves the use of sensors, GPS, and data analytics.

- IoT and Sensor Networks: Data on soil moisture, temperature, and crop health is collected via IoT devices and sensors. This data is then analysed with artificial intelligence and machine learning algorithms to deliver actionable insights for accurate resource management.
- ICT-driven irrigation systems enable farmers to provide the correct amount of water to crops, decreasing water waste and energy expenditures. This is especially important in water-stressed states like Maharashtra and Gujarat.
- Drones and Satellite Imaging: Drones with cameras and multispectral sensors take high-resolution photos of fields. These photos and satellite data, assist farmers in monitoring crop conditions, identifying insect infestations, and precisely plan treatments.

3. ICT-Enabled Agri Startups

India's startup ecosystem has seen a surge in agri-tech startups leveraging ICT to address agricultural challenges.

- Farm Management Apps: Startups such as AgriBazaar and FarmERPconnect with other farmers to markets, provide real-time pricing information, and promote frictionless transactions. This empowers farmers to make informed decisions about when, where, and how to sell their produce.
- Crop Advisory Services: Companies like Cropln and RML AgTech use mobile apps to provide farmers personalized guidance. These recommendations are based on data analytics, weather forecasts, and soil health evaluations, and they assist farmers in optimizing crop yields and resource utilisation.
- E-commerce Platforms: Agri-tech businesses such as Ninjacart and BigHaat connect farmers directly to merchants and consumers, decreasing the role of intermediaries. Farmers benefit from higher pricing, while customers benefit from fresher produce.
- Supply Chain Optimisation: Startups are coming up with ICT solutions to improve supply chain efficiency and reduce post-harvest losses.

Technologies such as cold chain logistics and blockchain-based traceability are deployed to preserve product quality and transparency along the supply chain.

Conclusion

ICT has a transformative impact on Indian agriculture, promoting sustainability and resilience. Farmers may receive real-time data for effective resource management and greater productivity using precision agriculture, sensor networks, and drones. ICT promotes rural economic growth by facilitating market access, knowledge sharing, and sustainable practices. Initiatives like digital farming platforms and ICT-enabled companies connect farmers to critical resources and markets, accelerating agricultural modernisation. The use of ICT in India promises a data-driven, efficient, and sustainable future for the country's agriculture sector, benefiting both farmers and consumers.

References

Behera, B.S., Das, T.K., Jishnu, K.J., Behera, R.A., Behera, A.C. and Jena, S., (2015). E-Governance Mediated Agriculture for Sustainable Life in India. Procedia Computer Science, 48(1): 623-629.

GoI. (2022). Directorate General of Civil Aviation. Retrived from Welcome to DigitalSky (dgca.gov.in).

Ministry of Agriculture and Farmers Welfare, Government of India. (2021). Impact of eNAM on Indian Agriculture. Retrieved from Agricultural Statistics at a Glance - 2021 (English version).pdf (dacnet.nic.in).

More, A. And Aslekar, A. (2022). Role of ICT & Fintech in Indian agriculture. In 2022 International Conference on Decision Aid Sciences and Applications (DASA). pp. 900-904.

Saidu, A., Clarkson, A.M., Adamu, S.H., Mohammed, M., and Jibo, I. (2017). Application of ICT in agriculture: Opportunities and challenges in developing countries. International Journal of Computer Science and Mathematical Theory, 3(1): 8-18.

Sennuga, S.O., Conway, J.S. and Sennuga, M.A. (2020). Impact of information and communication technologies (ICTS) on agricultural productivity among smallholder farmers: Evidence from Sub-Saharan African communities. International Journal of Agricultural Extension and Rural Development Studies, 7(1): 27-43.

Spielman, D., Lecoutere, E., Makhija, S., and Van Campenhout, B. (2021). Information and communications technology (ICT) and agricultural extension in developing countries. Annual review of resource economics, 13(1): 177-201.

Swaminathan, M. and Swaminathan, M.S. (2018). ICT and agriculture. CSI Transactions on ICT, 6(1): 227-229.

5

Gender Dynamics in Indian Agriculture: Roles, Challenges, and Opportunities

Biswajit Mallick[*1] *and Jyotishree Anshuman*[2], *Dorisa Narzary*[3] *and Priya Ranjan Mohanty*[4]

[1,2]*PhD Scholar, Department of Agricultural Extension Education, College of Agriculture, Odisha University of Agriculture and Technology, Bhubaneswar, Odisha, India.*
[3]*Research Scholar, Department of Agricultural Extension Education, Dr. Panjabrao Deshmukh Krishi Vidyapeeth , Akola, College of agriculture , Nagpur.*
[4]*Research scholar, Department of agriculture extension, Palli siksha Bhavana, Visva Bharati University, Shantiniketan.*
**Corresponding author's email: biswajitmallick9999@gmail.com*

Abstract

Agriculture is the primary source of food production for the survival of the human population. It supports millions of people in terms of livelihoods and the development of local and national economies. Women perform a crucial role by directly or indirectly involved in agriculture viz., producing, processing, and marketing agricultural produce. Other than agricultural activities, they significantly contribute to livestock rearing, fisheries, forestry, and other related activities. However, factors like lack of education, technology, inputs, and credit availability create a huge gender difference even in the 21st century, results the ineffectiveness of extension programs that address women's needs and priorities. Before understanding women empowerment, the concept of gender roles, gender-based discrimination, gender analysis frameworks, and gender empowerment strategies are also essential. This chapter also highlights the initiatives taken by the central government to empower and provide safety to women in society.

Keywords: *Agriculture, Extension Programs, Gender Roles, Gender Analysis Tools, Women Empowerment.*

Introduction

Agriculture plays a vital role in human societies and economies around the world. It is the prime food source for human populations, ensuring a reliable supply of nutritious food for the health and well-being of people everywhere. It is a major contributor to the economy of many countries, particularly in the developing world, where it provides livelihoods for millions of people and supports local and national economies. Agriculture is both impacted by and contributes to climate change. However, sustainable agriculture cultivation can also help to reduce the consequences of climate change by decreasing emissions of greenhouse gas and increasing resilience to extreme weather events.

Women in agriculture refer to women's participation and involvement in agricultural activities, including farming, livestock rearing, fisheries, forestry, and other related activities. Women in agriculture may be farmers, farm laborers, or work in other roles such as food processors, marketers, or traders. Women play a significant role in world agriculture, accounting for around 43% of the global agricultural labor force, according to the Food and Agriculture Organization (FAO) of the United Nations. India is often called an agrarian country because agriculture has been a major part of the Indian economy for centuries. In rural areas, a large portion of the population is engaged in agricultural activities. Women's involvement in agriculture can significantly impact food security and poverty reduction, as women tend to reinvest a higher proportion of their income in their families and communities. About 54.6% of the country's workforce was engaged in agriculture and allied activities in 2019-20, out of which almost half of the agricultural workforce, i.e., 42%, comprises women (NSSO). In the agricultural year 2015-16, about 68%, i.e., approximately 47 million of female agricultural workers in India were engaged in crop cultivation, followed by 22% (15 million) of females were engaged in livestock rearing and 10% (7 million) of female were engaged in fishing and aquaculture activities (NSSO). This is a significant figure and highlights women's crucial role in India's agriculture sector. Despite women's participation in almost all farming activities, their status in agriculture as a labourer is not highlighted in India. (Singh, 2013).

Role of Extension towards women development in Agriculture

Extension is a key component of agricultural development, as it involves disseminating information, knowledge, and skills to farmers and other stakeholders in the sector. In the context of women in agriculture, extension can play a crucial role in empowering women by providing them with the information and tools they need to increase their productivity, improve their livelihoods, and participate fully in decision-making processes.

The extension research system is having inbuilt component of gender. Still, it remains inattentive, so women often face challenges in accessing extension services, including limited access to information, lack of time due to household and caregiving responsibilities, gender-based discrimination, limited access to credit, and fewer opportunities to participate in farm decision-making processes (Unnati et al., 2012, Bala, 2010). Jemimah et al. (2013) reported that out of 923 sampled households, 38.3% of women had access to Agricultural inputs, followed by extension service (23.3%), agricultural loans (17.3%) and output markets (14.3%) in India. Even women farmers are just as productive and efficient as male farmers. Still, the major gender gap exists in the lack of access to resources like education, technology, inputs, and credit to their male counterparts in India. As a result, extension programs may not effectively address women's needs and priorities and may even perpetuate gender disparities in access to resources and opportunities. Hence, there is a need to recognize the importance of both genders in agriculture and initiatives to address the gender gap in access to resources and opportunities. This chapter highlights gender roles, gender-based discrimination, gender analysis frameworks, and gender empowerment strategies in the agricultural sector.

Important facts and figures about farm women in India

S. No.		**Overall**		**Rural**		**Urban**	
1.	Population of India	1210.9M		833.1M	68.8%	377.8M	31.2%
2.	Female	587.7M	48.47%	286.2	48.69%	301.5	51.31%
3.	Sex ratio	943		949		929	
4.	Literacy	65.46%		57.93%		79.11%	
5.	Female workforce	25.51%		30.02%		15.44%	
6.	Women are paid compared to men	-		60%		80%	
7.	Female cultivators	31.11%		-		-	
8.	Female in Agricultural research system	17.40%		-		-	
9.	Average wage per man day worked for women	145.63		-		-	
10.	Average daily wage for unskilled women labourer	131.09		-		-	

***Source*:** Census 2011

Important terminologies related to gender

Sex

It refers to the biological differences that divide males and females, including reproductive organs, hormones and chromosomes. It is typically assigned at birth based on physical attributes, but it's important to recognize that biological sex exists on a spectrum and can be more complex than a simple binary classification.

Gender

Gender is defined as the socially constructed differences in roles, responsibilities and identities related with being male or female in every society. Unlike biological sex, gender is a complex and fluid concept that varies across cultures and can encompass a range of identities beyond the traditional male and female binary.

Gender discrimination

Gender discrimination is the unfair treatment or exclusion of individuals based on gender, often resulting in unequal opportunities, rights, and privileges. It perpetuates harmful stereotypes and limits the full potential and contributions of individuals, hindering social progress and equality.

Gender mainstreaming

Gender mainstreaming is the integration of gender perspectives into all aspects of policies, programs, and initiatives to address gender inequalities and promote gender equality. It involves considering the different needs, experiences, and priorities of women, men, and gender-diverse individuals to ensure equitable outcomes and social inclusion.

Gender awareness

Gender awareness involves recognizing and understanding the social constructs, roles, and expectations associated with gender.

Gender sensitivity

Gender sensitivity is the ability to acknowledge and respect the diverse experiences, perspectives, and needs of individuals based on their gender, promoting inclusivity and equality.

Gender bias

Gender bias refers to the unfair or discriminatory treatment, attitudes, or stereotypes based on a person's gender, which can perpetuate inequalities and hinder opportunities.

Gender analysis

Gender analysis systematically examines how gender roles, relations, and power dynamics influence social, economic, and political processes, enabling a deeper understanding of gender inequalities and informing strategies for change.

Gender equity

Gender equity ensures fairness and justice by providing equal opportunities and resources to all genders, and addressing historical and structural inequalities.

Gender equality

Gender equality aims to eliminate discrimination and promote equal rights, opportunities, and treatment for all genders, creating a society where individuals can thrive regardless of gender identity or expression.

Gender blindness

Gender blindness refers to ignoring or disregarding gender differences and their impact, which can perpetuate inequalities and hinder efforts to achieve gender equality.

Gender roles

Gender roles are the societal expectations, norms, and behaviors associated with being male or female, influencing individuals' behaviors, roles, and responsibilities in various aspects of life.

Practical gender needs

Practical gender needs to address immediate challenges individuals face based on their gender roles, such as ensuring equal access to healthcare, education, and economic opportunities.

Strategic gender needs

Strategic gender needs to focus on addressing the root causes of gender inequality and transforming societal structures and norms to achieve long-term gender justice and empowerment.

Analysis tools and methods for gender analysis in agriculture

1. Gender Planning Framework

The Moser (Triple Roles) Framework, developed by Caroline Mosher, is a valuable tool used in gender analysis to understand the division of labor within households and communities. It emphasizes the concept of women's "triple role" in production (such as farming), reproduction (household chores and childcare), and community affairs, highlighting their simultaneous engagement in multiple roles. On the other hand, men are typically more involved in productive and community management activities while having less involvement in household tasks.

Above framework helps extension personnel to examine how women and men manage their roles and identifies specific agricultural extension approaches best suitable to their needs. As a result, it enables the analysis of gender roles regarding gender needs, distinguishing between practical needs that support daily life, such as obtaining extension information, and strategic needs that address gender disparities, such as literacy training.

2. Women Empowerment Framework

The Sara Longwe's Women Empowerment Framework (WEF), is a well-known conceptual framework that focuses on women's empowerment in the context of development programs and policies. The framework highlights five critical dimensions that need to be addressed to achieve women's empowerment:

i. Welfare: This dimension focuses on meeting women's basic needs and ensuring their physical well-being, such as access to healthcare, education, and nutrition.

ii. Access: It emphasizes women's equal access to resources, opportunities, and decision-making processes in various spheres, including education, employment, land, credit, and political participation.

iii. Conscientization: This dimension highlights the importance of raising women's awareness of their rights, capabilities, and societal roles, enabling them to challenge and overcome gender norms and stereotypes.

iv. Participation: It emphasizes the active participation of women in decision-making processes, both within their households and in the broader community and political arenas.

v. Control: This dimension focuses on women's control over resources, assets, and decision-making power, giving them autonomy and agency over their own lives.

3. The Gender Analysis Matrix (GAM)

The above one is developed by Rani Parker. It is is a participatory tool used to examine the differential impacts of agricultural practices on men and women within a community. Through community engagement, the GAM enables participants to explore and challenge gender roles, analyze culturally-defined practices, and identify areas of discrimination or inequality. For instance, the community can assess labor practices and their effects on the well-being, time allocation, and earning potential of men and women. If discriminatory practices are identified, the community can work towards creating a more equitable environment by addressing them.

The GAM offers several benefits, primarily through its community-based approach. It facilitates collecting and analyzing information on the spot, allowing the community to identify and propose culturally validated and equitable solutions. By using a participatory approach, involving both men and women or separate gender groups, the GAM empowers the community to drive transformative change at the local level.

4. Social Relations Framework

The Social Relations framework for gender analysis, developed by Naila Kabeer at the Institute of Development Studies, Sussex University, UK, in collaboration with various stakeholders, provides a comprehensive approach to understanding gender inequality. The framework is rooted in key theoretical concepts, starting with the notion of development as the enhancement of human well-being. It also emphasizes the significance of social relations, the structural relationships that create and perpetuate systematic differences among different groups of people. Institutional analysis is another crucial aspect of the framework, as it recognizes that gender inequality is not limited to the micro-level of households but is reproduced across institutions at various levels.

Kabeer challenges the assumption of ideological neutrality and independence of institutions, highlighting how they produce and reinforce social differences and inequalities. Institutions are interconnected and operate at macro, meso, and micro levels, with changes in national institutions potentially impacting institutions at other levels. While institutions may vary across contexts and cultures, they share common aspects such as rules, resources, people, activities, and power. These dimensions are vital for analyzing social and gender inequality, as they help identify who performs certain roles, who benefits, and who experiences losses among men and women.

By examining institutions based on their rules, practices, people, resource distribution, and power structures, the framework enables a deeper understanding of gender inequality and its effects. It provides a valuable tool for analyzing social relations and identifying strategies for promoting gender equality.

5. Harvard Analytical Framework

The Harvard Analytical Framework, also known as the "Gender Roles Framework," is a practical toolset used to identify and analyze the division of labor between men and women in households, farms, or communities. It also helps document gendered differences in access and control over resources such as land, water, seeds, or extension information. Extension agents can utilize the information gathered from this toolset better to understand the roles and responsibilities of men and women, allowing them to tailor programs and interventions to meet the specific needs and interests of farmers or community members, particularly women. This toolset can be easily adapted for agricultural, livestock, or combined production systems.

To ensure the effectiveness of the toolset, it is crucial to collect detailed data at the individual, household, or community level. Most recommended approach for data collection is to conduct meetings by using participatory method with farmer respondents or community members. These meetings provide an inclusive space for everyone to contribute to the discussion, help map out the division of labor, and understand who is responsible for what, when, and how. If necessary, separate meetings can be held for men and women to ensure comfort and encourage participation, with the findings later shared in joint sessions. The data collection process involves using three interconnected tools: an activity profile, an access and control profile, and a list of influencing factors. Each tool includes a series of checklists featuring key questions that extension agents can employ during their assessments.

Central government sponsored schemes for women empowerment

1. Working Women Hostel

Launch Year: 1972

Aim: To provide safe and affordable accommodation for working women.

Objectives: Offering secure and convenient housing options for working women, promoting their safety, well-being, and independence and facilitating their access to employment opportunities.

2. Self Help Group (SHG)

Launch Year: The SHG concept has been implemented in India since the 1980s.

Aim: To promote the socio-economic empowerment of women through self-help and collective action.

Objectives: Facilitating group formation among women, providing access to financial services, building capacity, and skills, promoting entrepreneurship and income-generation activities, fostering savings and credit facilities, enhancing decision-making power, and improving the overall well-being of women.

3. Rashtriya Mahila Kosh

Launch Year: 1993

Aim: To provide micro-credit and financial services to women in need.

Objectives: Supporting women's economic empowerment through access to credit, promoting entrepreneurship, enabling income-generation activities, and improving women's financial inclusion and self-reliance.

4. Women Helpline Scheme

Launch Year: 1998

Aim: To provide a 24-hour toll-free helpline for women in distress.

Objectives: Offering counseling, information, crisis intervention, and referral services to women facing violence, harassment, or other difficulties, ensuring timely help and support.

5. Swadhar Greh Scheme

Launch Year: 2002

Aim: To provide support and rehabilitation services to women in difficult circumstances.

Objectives: Offering shelter, counseling, medical aid, legal support, and rehabilitation services to distressed women, including victims of violence, trafficking, and homeless women.

6. Rajiv Gandhi National Creche Scheme

Launch Year: 2006

Aim: To provide daycare facilities for children of working mothers.

Objectives: Ensuring a safe and nurturing environment for children, enabling working women to balance their work and family responsibilities, and promoting child development and early education.

7. National Mission for Empowerment for Women

Launch Year: 2010

Aim: To strengthen women's overall development and empowerment.

Objectives: Promoting gender equality, enhancing women's social, economic, and political empowerment, addressing violence against women, improving health and education outcomes for women, and creating an enabling environment for women's empowerment.

8. National Rural Livelihood Mission (NRLM)

Launch Year: 2011

Aim: To alleviate rural poverty and empower women by forming self-help groups (SHGs) and their federations.

Objectives: Enhancing women's social and economic empowerment, promoting their active participation in decision-making processes, improving livelihood opportunities through skill development and entrepreneurship, strengthening women-led grassroots institutions, fostering gender equality, and creating sustainable pathways out of poverty.

9. e-Samvad Portal:

Launch Year: 2014

Aim: To provide a platform for communication and exchange of information between the Ministry of Women and Child Development and the states/union territories.

Objectives: Facilitating effective communication, collaboration, and coordination between the central and state governments for women and child welfare initiatives, policies, and programs.

10. Beti Bachao, Beti Padhao (Save Daughters, Educate Daughters)

Launch Year: 2015

Aim: To address the declining child sex ratio and promote the education and welfare of girls.

Objectives: Preventing gender-biased sex selection, ensuring the survival and protection of the girl child, eliminating gender-based discrimination, promoting girls' education and their participation in all spheres of life, and creating an enabling environment for their holistic development.

11. One-Stop Centre

Launch Year: 2015

Aim: To provide integrated support and assistance to women affected by violence under one roof.

Objectives: Ensuring immediate access to support services for survivors of violence, facilitating their safety, providing counseling, legal aid, medical support, and rehabilitation services, and promoting gender-sensitive support mechanisms.

12. Sukanya Samriddhi Yojana

Launch Year: 2015

Aim: To promote long-term savings for the girl child and provide financial security for her future education and marriage expenses.

Objectives: Encouraging parents to invest in their daughters' education and overall development, fostering a sense of financial security and independence for girls, ensuring equal opportunities for girls in education and career choices, and challenging gender stereotypes and discrimination.

13. Mahila e-Haat

Launch Year: 2016

Aim: To empower women entrepreneurs by providing them a platform to showcase and sell their products online.

Objectives: Promoting economic empowerment of women, fostering entrepreneurship and self-reliance, facilitating market access and visibility for women-led businesses, encouraging skill development and innovation, and promoting gender-inclusive economic growth.

14. Mahila Police Volunteer

Launch Year: 2016

Aim: To enhance the safety and security of women in the community.

Objectives: Mobilizing trained female volunteers to act as a link between the police and the community, facilitating the reporting of crimes against women, raising awareness on women's rights, and providing support and assistance to women in need.

15. Ujjwala Yojana

Launch Year: 2016

Aim: To provide free LPG connections to women from below-poverty-line households and promote clean cooking fuel.

Objectives: Reducing indoor air pollution-related health hazards, preventing respiratory diseases, improving women's safety and convenience, reducing the time and effort spent on traditional cooking methods, enabling women to engage in productive activities, and promoting sustainable energy practices.

16. Maternity Benefit Programme

Launch Year: 2017

Aim: To provide financial assistance to pregnant and lactating women for improved health and nutrition during motherhood.

Objectives: Reducing maternal and infant mortality rates, promoting safe motherhood practices, ensuring access to prenatal and postnatal care, and enhancing the well-being of mothers and children.

17. PM Mahila Shakti Kendra Scheme

Launch Year: 2017

Aim: To empower rural women through community participation and development.

Objectives: Promoting women's entrepreneurship, supporting skill development and capacity-building, creating awareness of women's rights and entitlements, facilitating access to government schemes and services, and fostering women-led community development.

18. Pradhan Mantri Matru Vandana Yojana (PMMVY)

Launch Year: 2017

Aim: To provide financial assistance and support to pregnant and lactating women for improved health and nutrition during motherhood.

Objectives: Reducing maternal and infant mortality rates, promoting proper antenatal and postnatal care, encouraging good nutrition and health practices among pregnant and lactating women, enhancing their well-being and that of their children, and ensuring their overall empowerment.

19. SHE-BOX (Sexual Harassment Electronic-Box) Portal

Launch Year: 2017

Aim: To provide a mechanism for registering and addressing complaints of sexual harassment at the workplace.

Objectives: Ensuring a safe and secure working environment for women, facilitating the reporting of workplace harassment incidents, enabling timely redressal and action against offenders, and promoting a harassment-free workplace culture.

20. Support to Training and Empowerment Programme

Launch Year: 2017

Aim: To enhance the employability and skills of women through training and capacity-building initiatives.

Objectives: Providing vocational training, skill development, entrepreneurship training, and capacity-building opportunities for women, enabling their economic empowerment and sustainable livelihoods.

21. Nand Ghar Yojana

Launch Year: 2018

Aim: To provide quality early childhood care, education services, healthcare facilities for children, and nursing care support for mothers.

Objectives: Promoting the holistic development of children, ensuring proper nutrition and healthcare, supporting working mothers, and fostering early learning and cognitive development.

22. NARI Web Portal

Launch Year: 2018

Aim: To serve as a one-stop platform for information, resources, and services related to women's empowerment and welfare.

Objectives: Providing information on government schemes, policies, and initiatives for women, facilitating access to support services, promoting awareness of women's rights and entitlements, and fostering gender equality.

Conclusion

Farm women empowerment is getting importance after reorganization of gender role in agriculture. Gender issues are imperative to focus because gender inequality leads to affect farm production and productivity, access and control over resources and opportunities, quick decision making. Various gender analysis tools are also help to calculate opportunity and challenges related to both men and women participation in agriculture. Major and highlighted government initiative programmes empowers women for a safety and healthy life.

References

Acker, J. (1990). Hierarchies, Jobs, Bodies: A Theory of Gendered Organizations. Gender & Society, 4(2) (pp. 139-158).

Bala, N. (2010). Selective discrimination against women in Indian agriculture: A review. Agricultural Reviews 31 (3) (pp. 224-228).

Connell, R. W. (2009). Gender: In World perspective (2nd ed.). Polity Press.

Goulding, K. (2013). Gender dimensions of national employment policies. ILO .

Jemimah, N., Elizabeth, K. and Laurie, S. (2013). Increasing the productivity and empowerment of women smallholder farmers. CARE Pathways to Empowerment Report.

Kabeer, N. (1994). Reversed Realities: Gender Hierarchies in Development Thought. Verso.

Kabeer, N. (2005). Gender Mainstreaming in Poverty Eradication and the Millennium Development Goals: A Handbook for Policy-Makers and Other Stakeholders. Common wealth Secretariat.

March, C., Smyth, I., & Mukhopadhyay, M. (1999). A Guide to Gender-Analysis Frameworks. Oxfam.URL:www.ndi.org/files/Guide%20to%20Gender%20Analysis%20Frameworks.pdf.

Ministry of Finance, Government of India. (n.d.). Sukanya Samriddhi Yojana. Retrieved from https://www.nsiindia.gov.in/sukanya-samriddhi-yojana

Ministry of Petroleum and Natural Gas, Government of India. (n.d.). Pradhan Mantri Ujjwala Yojana. Retrieved from https://pmuy.gov.in/

Ministry of Rural Development, Government of India. (n.d.). National Rural Livelihood Mission. Retrieved from https://nrlm.gov.in/about-us/nrlm

Ministry of Women and Child Development, Government of India. (2018). Nand Ghar Yojana. Retrieved from https://wcd.nic.in/schemes/nand-ghar-yojana

Ministry of Women and Child Development, Government of India. (2017). Maternity Benefit Programme. Retrieved from http://www.wcd.nic.in/schemes/maternity-benefit-programme-mbp

Ministry of Women and Child Development, Government of India. (2018). NARI Portal. Retrieved from http://nari.nic.in/

Ministry of Women and Child Development, Government of India. (2017). SHE-BOX (Sexual Harassment Electronic-Box). Retrieved from http://shebox.nic.in/

Ministry of Women and Child Development, Government of India. (2015). One Stop Centre Scheme: Guidelines for Implementation. Retrieved from https://wcd.nic.in/sites/default/files/OSC_Guidelines_0.pdf

Ministry of Women and Child Development, Government of India. (n.d.). Rashtriya Mahila Kosh. Retrieved from http://www.rmk.gov.in/

Ministry of Women and Child Development, Government of India. (n.d.). Beti Bachao, Beti Padhao. Retrieved from https://www.wcd.nic.in/schemes/beti-bachao-beti-padhao

Ministry of Women and Child Development, Government of India. (n.d.). Pradhan Mantri Matru Vandana Yojana. Retrieved from https://www.wcd.nic.in/schemes/pradhan-mantri-matru-vandana-yojana

Ministry of Women and Child Development, Government of India. (n.d.). Mahila e-Haat. Retrieved from https://www.wcd.nic.in/schemes/mahila-e-haat

Moser, C. (1993). Gender Planning and Development: Theory, Practice, and Training. Routledge.

Moser, C. O. (1993). Gender Planning and Development: Theory, Practice, and Training. London: Routledge.

NSSO. (2010). Situation assessment survey of farmers. National Sample Survey Organisation, Ministry of Statistics and Programme Implementation, Government of India.

Ridgeway, C. L. (2011). Framed by Gender: How Gender Inequality Persists in the Modern World. Oxford University Press.

Riger, S. (1992). Epistemological Debates, Feminist Voices: Science, Social Values, and the Study of Women. Signs: Journal of Women in Culture and Society, 18(1) (pp. 1-63).

Singh, V. (2013). Gender participation in Indian agriculture: An ergonomic evaluation of occupational hazard of farm and allied activities. International Journal of Agriculture, Environment & Biotechnology (1) (pp. 157-168).

Smyth, I., & Mukhopadhyay, M. (1999). A Guide to Gender-Analysis Frameworks. Oxfam. Retrieved from Oxford. www.ndi.org/files/Guide%20to%20Gender%20Analysis%20Frameworks.pdf

Swim, J. K., & Cohen, L. L. (1997). Overt, Covert, and Subtle Sexism: A Comparison Between the Attitudes Toward Women and Modern Sexism Scales. Psychology of Women Quarterly, 21(1) (pp. 103-118).

The Global Development Research Center. (n.d.). Gender Analysis Framework. Retrieved from www.gdrc.org/gender/framework/framework.html

United Nations Development Programme (UNDP). (1995). Human Development Report 1995: Gender and Human Development. Oxford University Press.

United Nations. (1979). Convention on the Elimination of All Forms of Discrimination against Women.

United Nations. (2015). Transforming Our World: The 2030 Agenda for Sustainable Development.

Unnati, A., Ankush, G. S. and Mande, A. V. (2012). Extent of participation of farm women in decision making. Journal of Dairying Foods & Home Sciences 31 (1) (pp. 72-74).

World Bank. (2001). Engendering Development: Through Gender Equality in Rights, Resources, and Voice. Oxford University Press.

6

Critical Study on Farmers' Producer Organization in India

*Netrananda Das and Satarupa Modak**

PG Scholar and Assistant Professor, Department of Agricultural Extension Education, M S Swaminathan School of Agriculture, Centurion University of Technology and Management, Odisha, India -761211.
**Corresponding author's email:* satarupamodak0@gmail.com

Abstract

Farmer Producer Organizations (FPOs) have become important players in the agricultural industry, supported by favourable legal and policy frameworks and government programmes that aim to give smallholder farmers more authority. By increasing farmers' income and access to resources and so reducing poverty, these organizations have demonstrated a major socioeconomic impact. Despite obstacles such limited finance and managerial difficulties, FPOs are crucial in promoting contemporary and sustainable agriculture practises. Additionally, by advocating environmentally friendly farming methods, they help to advance sustainable agriculture and the acceptance of climate change. FPOs are poised to continue growing and being resilient in the future, making them important players in agriculture, especially in light of changing climatic conditions and market dynamics.

Keywords: *Farmer producer organization (FPO), small and marginal farmers, Poverty Alleviation.*

Introduction

The Indian economy depends heavily on agriculture, which is also key to its expansion and advancement. In India, agriculture and related industries support more than two thirds of the people either directly or indirectly. 85 percent of Indian agriculture's growers are small and marginal farmers (GOI, 2010-11). It is difficult for them to access markets, knowledge, loans, and agricultural extension services. As a result of the high transaction costs, the food sectors, finds challenges to buy agricultural products directly from farmers. This renders the supply chains ineffective and raises issues with food safety, lack of transparency, and traceability. Although debt is frequently stated as the direct cause of their misery, deeper problems including low scale operations, a lack of information, inadequate communication links with the wider market, etc., are also involve in these phenomena. However, the daily rise in suicide rates in agriculture demonstrates how difficult it is for these farmers to survive. Due to the limited negotiating power, small landowners are more dependent on monopolistic exploitation during the cultivation process (Bachke, 2009). Though, Indian agriculture has a magnificent growth over last few years that the production of foodgrains have increased from 51 million tons to 314 million tones over past few years from 1950-51 to 2021-22 but has many claws.In addition to these related concerns, farmers also encounter tremendous exploitation from middlemen while obtaining inputs and marketing their produce.However, the agricultural sector has seen numerous problems that must be resolved in order to achieve sustained economic development (Mariappan and Zhou, 2019).

By commercializing and diversifying the agricultural venture of a farmer can help to improve their status.It is generally known that small-scale, resource-constrained farmers' commercialization leads to increased specialization, higher production, and higher income (Bernard and Spielman, 2009). The grouping of marginal and small farmers into farmer producer groups has emerged as one of the most effective ways to address agricultural issues, particularly in terms of greater access to investments, technology, inputs, and markets.While production is mostly left to individual small farms, the strategy is to integrate the growers directly into the market through their institutions (producer companies/cooperatives), marketing both output and inputs through collective processing and marketing. Farmer producer organizations registered under the special provisions of the Companies Act, 1956 have been identified by the Department of Agriculture and Cooperation, Ministry of Agriculture, Government of India, as the most effective institutional form around which to organize farmers and increase their capacity to pool their production and marketing power.

In order to revitalize agriculture, this situation calls for significant structural reforms and transformational initiatives. These reforms should be made in agricultural marketing and post-harvest logistics, as well as by increasing investments for productivity enhancement. Organizational designs for integrating Farmers into the value chain are emerging. Farmer Producer Organizations (FPOs) are one such project that aims to assist farmers in gaining the advantages of economies of scale through collectivization and aggregation.

Farmers' Producer Organization (FPO) is an independently owned company run by farmers. An effort is being made by the central government to help farmers overcome the various challenges that small producers are currently facing.FPO is an approach to bring together small and marginal farmers with other small producers to launch their own business under the guidance of professionals.FPO may help farmers market their crops as well as produce a number of agricultural goods. Small farmers are assisted by farmer organizations to engage in the market more successfully and cooperatively. As a result, their access to inputs and outputs, collecting the essential market knowledge, gaining access to new technologies, and participating in high value markets is made more affordable, allowing them to compete with larger farms and agribusinesses.

According to some literaturesreview FPOs are farming organizations that seek to strengthen the financial position of farmers by offering comprehensive services and assistance. By removing middlemen from the agricultural marketing process, FPOs enable farmers to get a higher share of the value that consumers are willing to pay (Chandrakar et al., 2023). According to research conducted in the Banda District of the Bundelkhand region, FPO members who grew chickpeas experienced higher net returns and lower cultivation costs than non-members (Rai et al., 2023). FPOs are nonprofit organizations that actively involve farmers in decision-making without making any distinctions on the basis of gender, socioeconomic background, or religion (Falguni et al., 2022). Another study in Andhra Pradesh's Guntur District found that rice farmers working under FPOs had great technical efficiency, with the potential to lower input by 1.3% while keeping output at the same level (Sanghamitra & Kumar, 2023).

Description of the topic:

Evolution and Growth of FPOs

Early initiatives and milestones

Farmers' Producer Organizations (FPOs) have been crucial in changing agricultural landscapes and giving farmers authority across the globe. In the early 2000s, India was home to one of the first and most important projects. The "FPO Promotion and Development Scheme" was launched by the National Bank for Agriculture and Rural Development (NABARD) in 2004 as a means of providing financial support for the promotion of FPOs throughout India and also the creation of the "Small Farmers' Agri-Business Consortium (SFAC)" in 2004 under the Ministry of Agriculture and Farmers' Welfare was one of the early attempts in this respect. By offering financial support, capacity building, and technical help to farmer groups, SFAC played a crucial part in supporting FPOs. This was a turning point for the FPO movement, giving them the funding, they desperately needed to start and flourish.

Another important turning point was the publication of the "National Policy for Farmers" in 2006, which acknowledged FPOs as a crucial element for the sustainable growth of agriculture and also the 2007 publication of guidelines for the establishment and development of FPOs by India's Ministry of Agriculture and Farmers Welfare. These regulations provided a foundation for the creation of FPOs, encouraging farmers to coordinate their agricultural operations and strengthen their market position.The "Task Force on FPOs" was established in 2011 under the leadership of renowned agricultural scientist Dr. M.S. Swaminathan, and it made many policy recommendations to enhance FPOs. The "Small Farmers Agribusiness Consortium (SFAC) Act" was passed in 2014 in order to give FPOs legal legitimacy and make the registration process simpler.

Current landscape and trends

Farmers' Producer Organizations (FPOs) are currently undergoing extraordinary growth and change in India. A number of government initiatives, such as the "Formation and Promotion of 10,000 FPOs"programme under the Atma Nirbhar Bharat Abhiyan, have speed up the development of FPOs throughout the nation. FPOs were gaining popularity and were being recognized as a potent tool for improving the income and way of life of small and marginal farmers. Government initiatives and financial institutions kept offering grants, training, and lending access to FPOs. In order to broaden their influence, FPOs were extending their reach beyond agricultural agriculture to include livestock, fisheries, and non-farm enterprises. The use of digital platforms by FPOs for marketing, e-commerce, and information sharing was a key contribution made by technological breakthroughs. FPOs were able to access new markets and strengthen their negotiating position because to this digital change.

In line with rising customer demand for environmentally friendly and wholesome agricultural products, FPOs were also boosting their use of sustainable and organic farming methods. As FPOs sought to take a bigger piece of the value chain, the movement towards value addition and Agro-processing was also gaining ground.

Legal and Policy framework

The development, operation, and expansion of Farmer Producer Organizations (FPOs) in India are significantly influenced by the regulatory environment. The regulatory environment for FPOs in India is primarily governed by a variety of legal and regulatory frameworks at the central and state levels.

- Companies Act, 2013: Under Section 581A of the Companies Act, 1956 (now the Companies Act, 2013), FPOs may register as Farmer Producer Companies (FPCs). FPOs are permitted to operate as businesses under this legal framework, which includes regulations unique to the agriculture industry.
- Cooperative Societies Act: FPOs can register as cooperative societies under state-specific Cooperative Societies Acts in addition to the Companies Act. This choice offers FPOs a more conventional cooperative structure.
- Multi-State Cooperative Societies Act, 2002: Under this legislation, FPOs have the opportunity to apply to become Multi-State Cooperative Societies, which enables them to conduct business in various states.

- NABARD (Amendment) Act, 2017: With the amendment of the National Bank for Agriculture and Rural Development (NABARD) Act, FPOs now have access to funding, credit options, and capacity-building support.
- National Policy for Farmers (2007): FPOs are significant in enhancing farmers' economic circumstances, according to the National Policy for Farmers. The government's promotion and support of FPOs are set in this document's direction.

Government initiatives and schemes:

In order to assist the establishment and development of Farmer Producer Organizations (FPOs) in India, the government has developed a number of initiatives and programmes. These programmes seek to enhance farmer lives, boost the agricultural industry, and encourage collective farming.

- Formation and Promotion of Farmer Producer Organizations (FPO) Scheme: The Indian government has introduced programmes to aid in the establishment, promotion, and capacity-building of FPOs under the Department of Agriculture, Cooperation & Farmers Welfare (DAC&FW).
- NABARD's Producer Organization Development Fund (PODF): The PODF was established by the National Bank for Agriculture and Rural Development (NABARD) to offer financial support and equity assistance to FPOs. Their capital basis and operating capacity are strengthened as a result.
- SFAC's Equity Grant and Credit Guarantee Fund (CGF) Scheme: To assist FPOs in accessing capital and reducing risk, the Small Farmers' Agribusiness Consortium (SFAC) offers equity grants and credit guarantee facilities.
- National Cooperative Development Corporation (NCDC): For a variety of tasks like marketing, processing, and infrastructure development, NCDC offers financial assistance to FPOs in the form of loans and grants.
- Pradhan Mantri Kisan SAMPADA Yojana: This programme, which is funded by the MoFPI, encourages the development of a food processing infrastructure that can help FPOs by supplying processing resources and lowering post-harvest losses.
- National Agricultural Market (e-NAM): Agricultural markets all around the country are electronically connected via the e-NAM project. This platform gives FPOs access to a larger client base and allows them to get higher rates for their produce.
- Price Support Scheme (PSS): FPOs are eligible to take part in PSS, a programme where the government purchases crops at Minimum Support Price (MSP) levels throughout market turbulence to guarantee farmers receive fair prices.

Case studies

Case study-1

Mahashakti Mahila Milk Producer Company Limited (2015)

Narayanpur Village, Thanagazi Block, Alwar District, Rajasthan

Product: Milk

In Rajasthan, a less developed area is the Thanagazi block in the Alwar district. In particular, women have extremely low literacy rates. The main occupation is agriculture. Agriculture and farmers, however, are now less productive due to a shortage of irrigation systems and water and also, animal husbandry has historically been a significant source of income, with women performing the majority of the work. It was difficult to market milk in a rural location at a price that was competitive. In the past, middlemen were in charge of collection. They used to buy milk below market value and did not take other aspects into account when setting prices, such as fat content. Woman farmers were also caught in a debt trap because of their low income and lack of alternative options. For that purpose, NSU came forward to look after these problems. Smt. Uma Ratnu (President) formed Nari UtthanSansthan (NSU), a grass-roots organisation with a presence in Rajasthan that has been working in the fields of women's empowerment, health, education, and awareness since 1995.

The creation of the FPO has benefited the women members in various ways. Members receive an improved and guaranteed price for their milk. Due to the guaranteed marketing facilities, many individual members are also utilising bank financing to scale up the dairy industry. Women members have escaped the middlemen's debt trap. They oversee all business operations, including collecting, processing, and marketing. Their abilities, leadership abilities, self-confidence, and self-esteem have all improved as a result. Members' net annual income ranges between Rs. 15,000 and Rs. 20,000. The net annual savings to each FPO member has often ranged between Rs. 5,000 and Rs. 6,000. The ladies who make up the Maha Shakti Milk Producer Company were already connected to a bank under NABARD's SHG scheme before they established their FPO. The FPO can obtain a basic credit facility thanks to their background and experience. Maha Shakti Milk Producer Company overcame the first difficulty faced by the majority of newly formed FPOs in obtaining finance with the assistance of Nari UtthanSansthan. The FPO established collection centres and processing equipment, each with a measuring cup/weighing machine. The Mahashakti FPO's milk processing unit has incorporated cost-effective technologies. From an operational perspective, there are no technical complications involved. They handle a variety of dairy products, including ghee, cheese, butter, and curds. As per 2017-18 FY it covered over 109 villages with a total active member of 778 and generated a total turnover of INR 150 crore.

The main activities of this FPO include training on the cultivation of vegetables, horticultural plants and spices; processing, grinding and packaging of agricultural products; training, exposure and capacity building of farmers; training on how to make ghee, mawa, milk cakes, etc.; community mobilisation and training for better farming; agricultural input supply to the shareholders (seeds, fertilisers); and marketing linkage initiatives for agricultural products.

Case study-2

Jaivik SRI Farmers Producer Company Ltd. (2016)

Koraput, Odisha

Product: Agricultural & NTFP Products

One of the main tribal districts in Odisha is the Koraput district. Its headquarters are located in a remote area of Odisha, south of the state's capital. Although many areas of the district have been gradually cleared for the exploitation of natural resources over the last few decades, it is located at a high elevation and has substantial forest cover. The region naturally has a high level of organic carbon in the soil, and the majority of the tribal farmers use organic farming practises. Industrial fertilisers and insecticides have gradually made their way into agricultural practises. The majority of the tribal population, as well as members of other communities like scheduled castes, have relied mostly on agriculture and the harvesting of forest products for their subsistence. Modern industrial agricultural inputs, including as fertilisers, pesticides, heavy machinery, etc., have also been introduced in the district thanks to government departmental involvement. The district's natural environment has become somewhat unbalanced as a result of the interaction of industrial inputs, policy, credit, trade, business, and people.

PRAGATI and NABARD worked together to implement a project on productivity increase of three important crops (Paddy, Pulses, Vegetables, and NTFP) of the Koraput area, with the subject of agro-based livelihoods as the main focus. In order to strengthen the agro-based livelihoods of the farming community in the project area, village level institutions like Self Help Groups (SHGs) and Farmers Producer Company were supported. This was one of the project's objectives. Farmers underwent agricultural training, learning first-hand about new varieties, effective organic agriculture, the use of organic fertilisers, and boiling devices.

Harsha Trust, one of the state's top nongovernmental organisations in this field, first promoted Jaivik Shri. But because it was unable to manage it, the task was assigned to Pragati, which with the assistance of NABARD worked in the communities on programmes to improve productivity and manage land and water. On May 18th, 2016, the FPO was officially registered. There are currently 37 villages represented by 13 GPs and 3 Blocks among its members.

The FPO has generally had a positive effect on the members. Many of the tribal farmers have developed more eloquent and outspoken ways of expressing themselves through their physical appearance. They have gained experience working cooperatively to plan their organic crop and vegetable production as well as the marketing of their produce. Through Pragati's guidance, they have also received experience running an organisation (FPO). Additionally, the FPO has made investments in a few infrastructure projects for goat and poultry farming, as well as a marketing outlet, all on the same property. According to the POPI, the rise in members' net annual income due to the creation of the FPO ranged between INR 14,000 to 32,000. When compared to the net income that the majority of FPO in the nation makes, this result is very exceptional. Additionally, it states that members would have benefited from a variety of services made possible by the FPO to the tune of INR 10,000 year. As of March 2018, this FPO's equity with the tribal people was worth INR 100,000. The whole income for 2017 was INR 935,484, and it climbed to INR 1,229,604 in 2018.

In 2017 and 2018, the FPO reported minor profits of INR 3243 and 5331, respectively. Although beneficial, these operating profits are not large. The important feature of this FPO, however, has been the relatively very high net gains to members. The members' additional net incomes, as reported by the POPI, range from INR 13,000 to INR 32,000 annually. FPO primarily focuses on agriculture, horticulture, sericulture, livestock, and pisciculture. It has members in 551 total, 335 of them are actively involved, and it serves 37 villages.

Case study -3

Ramanar Millets Farmer Producer Company Limited (2016)

Tamil Nadu

Product: Millet

Livelihood Sustainability of Farmers through FPO

Prior to the establishment of Ramanar FPO in 2016, an NGO by the name of SEEDS had been collaborating with the local population since 1999. Since 1999, the Tamilnadu State Rural Livelihood Mission (TNSRLM) has supported SHGs in the area. The Ramanar Producer Company (FPO) was established with the participation of FIG and JLG leaders. 10 active farmer members who had previously participated voluntarily in different rural agricultural development initiatives were drawn from these community-based organisations. Currently, the FPO has 2943 members from 112 villages in 47 Gramme Panchayats (GPs) from three Blocks, namely Tiruchuli, Narikudi, and Kariyapatti blocks of Virudhunagar District, and another Block, namely Kamuthi block of Ramanathapuram District. The FPO's stated Mission is to "organise 5000 farmers as shareholders in 5 years and provide support on soil test, agricultural inputs, credit facilitation, price fixation, marketing and livelihood activities for its members." The FPO's stated

Vision is "To attain the livelihood sustainability to the farmers through traditional agriculture."

With 2943 members in 112 villages, the company has been active. Of these, 2107 are female members and 836 are male members. Members fall into the following categories: small farmers (1866), marginal farmers (694), landless/tenant farmers (226), and large farmers (157) who purchase their agricultural products such as pulses, cereals, chilly, coriander, and millets at the ideal price, with accurate weighing, minimal transportation, and immediate payment. Small producers have benefited from breaking away from local traders, who frequently had poor conditions of trade with them. Better crop pricing, accurate selling weighting, reduced transportation costs, and immediate payment have all been made available for the producer members through this procedure. Within three years, the amount of business increased ten times, from roughly INR 26 lakhs to INR 273 lakhs. Approximately INR 1,08,820 in profits have been generated in the three years from registration.

36,43,000 Indian rupees is the current paid-up share capital. It was able to secure an equity grant of INR 10,000 from SFAC, New Delhi, and as of 2019, its total assets were around INR 6,31,54,810. From 2016–17 to 2018–19, the FPO's annual financial turnover increased from Rs. 26,66,466 to Rs. 1,37,41,452 and subsequently to Rs. 2,66,47,670. The three years' profits were, respectively, Rs. 60,698, Rs. 13,18,670, and Rs. 7,30,420. The majority of the FPO's product line is made up of different millets and pulses. In accordance with the supply and demand from the members, it also deals with other agricultural products.

Impact assessment

Socio-economic impact of FPOs

Farmers' Producer Organizations (FPOs) have a significant socioeconomic impact in India since they were first established. By pooling resources, increasing negotiating power, and obtaining access to necessary support services, these organizations have significantly improved the lives of small and marginal farmers. FPOs have facilitated farmers' access to capital, technology, and training, enabling them to embrace cutting-edge and environmentally friendly farming practices. FPOs have helped members achieve better pricing by marketing their produce as a group, so lowering their exposure to market swings. As a result, farmers now earn more money and are more stable financially. By including women and underrepresented groups in decision-making and revenue-generating processes, FPOs have furthered social cohesion and gender fairness.

Role in poverty alleviation

FPOs (Financial Participatory Organizations) are crucial to the fight against poverty. They offer financial support to low-income people, including those who are struggling to make ends meet in rural areas and small company owners with limited operating capital (Geremewe, 2018).In India, Farmers' Producer Organizations (FPOs) are crucial to the reduction of poverty. FPOs enable small and marginal farmers to more effectively access resources, information, and markets by grouping them into collectives (Santhosha et al., 2022; Raj et al., 2022) and also provide afinancial service organization (FPOs) by a range of financial services, such as loans, deposits, savings accounts, money transfers, and insurance, enabling people to grow their enterprises and improve their economic circumstances (Vishwakarma and Mujjoo, 2023). These organizations offer farmers the chance to diversify their sources of income, allowing them to partake in high-value industries including Agro-processing, organic farming, and animal raising. FPOs help members increase their revenue by securing higher pricing and lowering post-harvest losses through collective selling of their produce. FPOs also make it easier for members to access government programmes, technology, and financing, which further enhances their members' financial security. FPOs aid farmers in developing their skills, gaining access to inputs, establishing businesses, and effectively marketing their produce, all of which result in increased incomes (Permadi, 2019). Furthermore, FPOs frequently place a high priority on social inclusion, making sure that marginalized and female farmers profit from their initiatives. This helps to close gaps and encourages a fairer distribution of income. FPOs are fundamental in helping rural households escape poverty by enhancing their resilience and future economic opportunities. Overall, FPOs help to double farmers' revenue by lowering production costs, increasing productivity, and enhancing involvement in the value chain (Padaliya et al., 2022).

Influence in agricultural practices

- **Adoption of Contemporary Methods**: FPOs encourage their members to use contemporary and sustainable agriculture methods. They support training and capacity-building initiatives that educate farmers about effective crop management, water-saving strategies, organic farming, and the application of cutting-edge technologies. This boosts production and lessens the impact on the environment.
- **Crop Diversity:** FPOs promote crop diversity by exposing farmers to novel, high-value crops. FPOs encourage farmers to plant a greater variety of crops beyond traditional staples by collectively analyzing market demand and negotiating higher pricing. The income can increase and the danger associated with monoculture can be decreased by diversification.
- **Access to quality inputs:** FPOs assist farmers in getting affordable access to high-quality seeds, fertilizer, insecticides, and other inputs. They frequently purchase inputs in bulk, enabling members to take advantage of economies of scale and bargain with suppliers for better terms. This guarantees that farmers may obtain the resources they need to meet their unique agricultural needs.

- **Crop management and pest control:** FPOs spread information about integrated pest management (IPM) and other environmentally friendly methods of pest control. FPOs assist in reducing the reliance on chemical pesticides, which can be detrimental to both the environment and human health, by boosting IPM and encouraging the use of biopesticides and natural predators.
- **Water management**: The promotion of effective water management techniques, such as rainwater collection and drip irrigation, is a responsibility of FPOs. They teach farmers how to conserve water, which is especially important in areas with a lack of water.
- **Organic farming**: A lot of FPOs work to promote organic farming methods. They aid farmers in obtaining organic certifications and gaining access to markets where organic produce is more highly valued.
- **Post harvest handling**: FPOs train farmers in effective post-harvest handling procedures to reduce losses and preserve food quality. This covers instruction in grading, packaging, and storage techniques.
- **Market linkage:** FPOs eliminate middlemen by connecting farmers directly to markets. This enables farmers to adapt their production by assisting them in better understanding market demands and quality standards.

Challenges and constrains

Financial sustainability:

Farmers' Producer Organizations' (FPOs) ability to maintain a positive financial position is essential to determining their long-term success and influence. FPOs frequently struggle to become financially viable, but a number of techniques can improve their sustainability. It's important for FPOs to diversify their revenue streams; instead of relying primarily on grants and subsidies, they should look for money from their main business operations, like collective marketing, value addition, and service supply. To guarantee that resources are used effectively, effective financial management, including open accounting and careful planning, is necessary. A key factor in enabling FPOs to invest in infrastructure, technology, and capacity-building is access to finance and financial services. Building strong connections with financial institutions, government programmes, and partners in the private sector can also help FPOs maintain their financial viability.Additionally, to maintain their resilience and long-term performance, FPOs must continuously evaluate and modify their business models to account for shifting market dynamics and member needs.

Capacity building and training needs

The efficient operation of Farmers' Producer Organizations (FPOs) in India depends on capacity development and training. Training programmes are crucial to enhancing the abilities of FPO members because they frequently come from diverse backgrounds and have varying degrees of agricultural experience. In order to ensure that FPOs run effectively and honestly, these programmes ought to address financial management, governance, leadership, and cooperative values. Productivity and product quality can be increased by receiving training in sustainable agriculture methods, utilising contemporary technologies, and managing post-harvest. To enable FPOs to negotiate better rates and reach markets, marketing and value chain training is essential. To manage risks, traverse complex regulatory frameworks, and obtain credit and financial services, FPOs also require assistance in expanding their capability.To improve the knowledge and skills of both leaders and members and ultimately contribute to the success and sustainability of these organizations, training programmes should be consistently given and tailored to the unique needs of FPOs.

Marketing linkage and price realization

Farmers' Producer Organizations (FPOs) in India succeed and remain economically viable in large part due to market linkage and price realization. FPOs are essential in connecting small and marginal farmers with markets directly, reducing their reliance on middlemen, and ensuring more equitable price realization. FPOs frequently have more negotiation strength thanks to collectively negotiating for their produce and pooling resources, enabling them to get competitive prices and reach wider markets. FPOs can also add value to their products and brand them, which improve their ability to recoup their costs. Furthermore, FPOs are better able to match their production with market demand, lowering the danger of surplus or undersupply, thanks to market data and analysis. These initiatives not only boost the earnings of FPO members but also help the agricultural sector as a whole by stabilizing prices, lowering waste, and enhancing supply chain effectiveness. FPOs must continually fortify their connections to the market, adjust to shifting market dynamics, and preserve quality standards to keep customers' trust in order to have the most influence possible.

Role of FPOs in sustainable Agriculture

Environmental sustainability initiatives

Indian Farmers' Producer Organizations (FPOs) have grown more aware of the value of environmental sustainability in recent years. These grassroots organizations, which are made up of small and marginal farmers, have launched a number of projects to solve environmental issues and advance agricultural output. Adopting sustainable farming methods including organic agriculture, integrated pest management, and agroforestry is one noteworthy project. FPOs advocate the use of less chemicals, water conservation, and biodiversity enhancement on farms and for save water and improve agricultural resilience in the face of climate change, FPOs have played a crucial role in promoting effective water management strategies including drip irrigation and rainwater collection. Additionally, these organizations run afforestation campaigns to reduce soil erosion and enhance soil quality.

FPOs frequently encourage its members to participate in non-farm activities, such as beekeeping and sericulture, in addition to on-farm practices. These activities not only diversify revenue sources but also aid in pollination and the preservation of crucial ecosystems. FPOs in India are playing a critical role in developing environmental sustainability in agriculture while enhancing the livelihoods of small-scale farmers by promoting sustainable agricultural practices and supporting responsible land and resource management. Their activities support long-term food security and economic stability in rural communities in addition to the environment.

Climate change adaptation

India's Farmers' Producer Organizations (FPOs) have acknowledged the pressing need for adaptation to the problems brought on by climate change. The agriculture sector's efforts to build climate resilience are being led by these collectives, which are made up of small and marginal farmers. FPOs are actively engaged in climate-smart agriculture practises, which include modifying crop calendars, choosing crop types that are tolerant to changing climates, and putting new water management strategies into practise. The transmission of climate knowledge and information to members is a crucial component of FPOs' climate adaption strategy. To assist farmers in making knowledgeable choices about when to sow and harvest their crops, they provide access to weather forecasts, early warning systems, and climate-smart agricultural practises. FPOs also offer training programmes and capacity-building initiatives to improve farmers' comprehension of the effects of climate change and available adaptation options.

Additionally, FPOs encourage the use of sustainable and resource-conserving technology to lower greenhouse gas emissions and boost overall resilience, such as drought-resistant seeds, precision farming equipment, and renewable energy sources. Additionally, they promote crop diversity to make sure farmers are not unduly reliant on a particular crop and to lessen sensitivity to dangers associated with the environment. And fundamentally, FPOs in India are actively involved in climate change adaptation initiatives, providing small-scale farmers with the information, resources, and techniques necessary to successfully deal with the problems presented by a changing climate and develop a more resilient and sustainable agricultural sector. In light of the unpredictability of the environment, their proactive approach is essential for preserving food security and the wellbeing of rural populations.

Future prospects

Potential for FPOs expansion

Farmers' Producer Organizations (FPOs) in India have a lot of potential to grow in the future. Although these collectives have already made significant advancements in empowering small and marginal farmers, their growth trajectory is yet optimistic. The future of FPOs in the nation is promising for a number of reasons. First off, the government's ongoing assistance in the form of regulations and financial incentives strengthens the contribution of FPOs to the transformation of the agricultural scene. Plans like "Formation and Promotion of 10,000 FPOs" have supplied much-needed funding and infrastructure, creating an atmosphere that is suitable to their growth. The second factor that will probably contribute to the rise of FPOs is greater farmer awareness and education about their advantages. Joining FPOs will become more appealing as more farmers see the benefits of increased negotiating power, better access to markets, and shared resources.Additionally, FPOs are adjusting to contemporary technologies and digital platforms in order to expand their audience and streamline their business operations. Farmers will find it simpler to interact with FPOs, get market data, and use digital payment systems because to this technical improvements. Last but not least, FPOs' objectives are in line with the urgent issues brought about by climate change and the demand for sustainable agricultural practices. FPOs can be essential in advancing environmentally friendly farming methods and resource-saving technology as climate resilience assumes greater importance.

Policy recommendation

Future policy suggestions for Farmer Producer Organizations (FPOs) should focus on enhancing and supporting these organizations to boost agricultural productivity, advance sustainable agricultural practices, and improve the livelihoods of small and marginal farmers. Here are some crucial directives for FPOs:

- Financial assistance and rewards: For operating capital requirements, infrastructure expansion, and capacity growth, provide FPOs with grants, subsidies, and low-interest loans. By offering tax advantages and incentives, promote both public and private sector investment in FPOs.
- Capacity building: For FPO members, offer training and skill-development courses with an emphasis on marketing, financial management, modern farming techniques, and cooperative governance. Encourage FPOs to use digital tools for better record-keeping, market access, and decision-making.
- Market access: You can make it simpler for FPOs to access local, regional, and global markets by partnering with agricultural businesses, exporters, and online retailers. Create farmer-friendly cold storage, processing facilities, and transportation infrastructure.

- Crop Diversification and Sustainable Practices: To increase soil health and lessen negative environmental effects, encourage FPOs to promote crop diversity and environmentally friendly farming methods including organic farming and integrated pest control. Offer rewards to FPOs engaged in the cultivation of crops with high value and resiliency to climate change.
- Risk Mitigation and Insurance: To safeguard farmers from crop failures and disasters caused by nature, develop and promote cost-effective crop insurance programmes specifically designed for FPOs. To provide early warning and reaction mechanisms, establish risk assessment and management units within FPOs.
- Research and Development (R&D): Invest money in crop types that are appropriate for each FPO's location and climate as well as agricultural research and extension services to fulfil their unique needs. To share information and transfer technology, encourage cooperation between FPOs and research institutes.
- Access to Technology and Innovation: To increase production and cut labour costs, encourage the creation and adoption of cutting-edge technologies like farm automation and precision agriculture instruments. To give FPOs access to cutting-edge solutions, facilitate cooperation with technology firms and start-ups.
- Social Inclusion and Gender Equity: By giving them training, funding, and representation in leadership roles, FPOs may assist the inclusion of marginalized and female farmers. Encourage public understanding of and instruction in gender-sensitive farming methods.
- Promotion of Agri-Entrepreneurship: Encourage FPO members to look at opportunities for Agri-entrepreneurship, such as food processing, value addition, and Agri-tourism, to boost income generation.

Conclusion

The analysis of Farmer Producer Organizations (FPOs) in India shows both their enormous potential and the difficulties they encounter. FPOs have shown promise as a means of empowering marginalized and small farmers, strengthening their capacity for collective bargaining, and raising agricultural output. They are crucial in bridging the gap between farmers and markets, improving rural livelihoods and earnings in the process. However, in order to realize their full potential, a number of problems must be resolved. These include the need for skill development among FPO members, regulatory barriers to registration, limited access to capital and technology, governance and management issues, and technological limitations. The dynamics of the market and the climate are also changing, and FPOs must adapt. In essence, while FPOs hold the key to changing India's agricultural environment, their success depends on aggressive government backing, the development of institutional capacity, and a comprehensive approach to rural development.

References

Bachke, M.E. (2009). Are farmers' organizations a good tool to improve small-scale farmers' welfare? Conference Paper Institute for Social and Economic Studies; Mozambique.

Bernard, T., & Spielman, D. J. (2009). Reaching the Rural Poor through Rural Producer Organizations? A Study of Agricultural Marketing Cooperatives in Ethiopia. Food Policy 34(1): 60-69. doi: https://doi.org/10.1016/j.foodpol.2008.08.001

Chandrakar, K., Chandrakar, D.K., Das, D., (2023). Farmers Producer Organization (FPO): How it is changing farming scenario. International Journal of Science and Research Archive, 08(01): 761–765. doi: https://doi.org/10.30574/ijsra.2023.8.1.0082

Falguni, R., Kumar, G.D.S. & Sharma, M.L. (2022). Development of a Scale to Measure Group Dynamics Effectiveness (GDE) of FPO Members. Indian Research Journal of Extension Education, 22(5): 146-149. doi: https://doi.org/10.54986/irjee/2022/dec_spl/146-149

Geremewe, Y.T. (2018). The Role of Micro and Small Enterprises for Poverty Alleviation. International Journal of Research Studies in Agricultural Sciences, 4(12): 38-47. doi: http://dx.doi.org/10.20431/2454-6224.0412005

Government of India (GoI). (2010-11). Agriculture Census Division, Department of Agriculture and Cooperation, Ministry of Agriculture, Government of India.

Mariappan, K. and Zhou, D. (2019). A threat of farmers' suicide and the opportunity in organic farming for sustainable agricultural development in India. Sustainability, 11(8): 2400. doi: https://doi.org/10.3390/su11082400

Padaliya, M., Devi, G., Raju, G.N.N.K. (2022). Farmer producer organizations way for enhancing rural livelihood. Gujarat Journal of Extension Education, 34(1): 90-96. doi: https://doi.org/10.56572/gjoee.2022.34.1.0018

Permadi, B. (2019). Community Empowerment and Farmer Poverty Reduction in Developing Countries. Journal of Public Administration Studies, 4(1). doi: https://doi.org/10.21776/ub.jpas.2019.004.01.2

Rai, R., Gupta, B., Kalia, A., Gautam, Y., Panday, H. (2023). Assessing the Comparative Economics of Chickpea: FPO in Banda District of Bundelkhand Region. Asian Journal of Agricultural Extension, Economics and Sociology, 41(5): 44-50. doi: http://dx.doi.org/10.9734/ajaees/2023/v41i51899

Raj, R., Perumal, T., Balaji, V. (2022). MobiMOOC – A Practical Learning Tool to Promote Corporate Literacy for Effective Functioning of Farmer Producer Organizations. doi: https://doi.org/10.56059/pcf10.2639

Sanghamitra, G. & Kumar, K.N.R. (2023). Analysis of Technical Efficiency of Rice Farmers under Farmer Producer Organization (FPO) in Guntur District of Andhra Pradesh. Asian Journal of Agricultural Extension, Economics and Sociology, 41(5): 127-134. doi: https://doi.org/10.9734/ajaees/2023/v41i51909

Santhosha, S.G., Anuradha, C.K., & Arunkumar, M. (2022). Conceptual study on role of farmer producer organization in measurement, evaluation and enhancement of sustainable agricultural effectiveness. International Journal of Health Sciences, doi: https://doi.org/10.53730/ijhs.v6nS6.9973

Vishwakarma, R.K. and Mujjoo, R., (2023), Role of microfinance in poverty alleviation in India. International Journal of Financial Management and Economics, 6(1): 221-226.

7

Concept of Diffusion and Adoption in Extension

Indira Priyadarsini Pattnaik* and Srabani Ghosal

PhD Scholar, Department of Agricultural Extension Education, Odisha University of Agriculture and Technology, Bhubaneswar, Odisha. Corresponding author's email: indirapattnaik143@gmail.com

Abstract

One of the primary objectives of extension services is to promote the adoption of new and profitable technologies within the farming community. Extension agents must possess a comprehensive understanding of adoption and diffusion processes to facilitate and expedite these processes effectively. In rural villages, it's evident that even when farmers are aware of innovative solutions, they don't always immediately embrace them. Instead, they often take some time to contemplate and make informed decisions. This phenomenon can be attributed to the diffusion process, which is the mechanism through which an innovation is transmitted over time within a social system via particular communication channels. At its core, the diffusion of innovations is known as a social process where individuals exchange their subjective perceptions and information regarding a new idea. It's crucial to acknowledge that the decision to adopt an innovation is usually not impulsive or immediate but rather a process of decision-making that unfolds gradually over time. During this process, individuals progress through various mental stages before arriving at a final decision to embrace the innovation. Extension agents can play a crucial role in guiding farmers through this decision-making process and providing the necessary information and support to encourage the adoption of beneficial agricultural innovations.

Key words: *Diffusion, Innovation, Invention, Social process, Social system*

Introduction

The primary mission of extension services is to facilitate desirable changes in human behaviour through educational means. These changes can encompass various aspects, including knowledge, skills, attitudes, understanding, goals, actions, and confidence. Effective communication is the cornerstone of an extension worker's role. Their responsibility extends beyond merely informing farmers about improved practices; they are tasked with ensuring that the results of research and field trials are practically applied by farmers. Nevertheless, extension workers often encounter perplexing questions in their line of work: Why does a disparity exist between what farmers know and what they put into practice? Where do most farmers acquire new ideas and practices? What factors contribute to some villagers swiftly adopting new practices while others are more hesitant? Why do certain farmers readily embrace and implement new ideas, while others are more cautious? What influences the time it takes for farmers to decide on adopting new ideas and practices? Why do some new ideas/practices gain acceptance effortlessly, while others require years of persistent efforts by extension agencies? Extension agents must possess a comprehensive understanding of adoption and diffusion processes. This knowledge equips them with the tools and insights needed to accelerate the adoption of agricultural innovations within their communities.

Diffusion process

Diffusion is the process through which an innovation gradually spreads among the members of a social system over time, utilizing specific communication channels (Ray, 2017). What distinguishes diffusion is the "newness" of the concept conveyed within the message, giving it a distinctive character. In essence, the diffusion of innovations is a social process where individuals exchange subjectively perceived information about a novel idea. Furthermore, this process introduces an element of uncertainty, leading to the acknowledgment of associated risks. It can be both a deliberate, planned process and a spontaneous one, depending on various factors and circumstances.

Components involved in the spread of innovations

The primary components encompass innovation, communication channels, the passage of time, and the social system.

1. Innovation

It is a concept, action, or entity that an individual or other adoption unit views as novel. Innovation and technology are often used interchangeably, with technology being essentially a blueprint for practical action that helps mitigate uncertainties in striving for a specific result by utilizing cause-and-effect relationships. Technology typically comprises two primary components: hardware and software, with the hardware component being more visible. The hardware aspect involves the physical tool that represents the technology as a tangible object. In contrast, the software aspect includes the informational basis for the tool (Rohilla, et al., 2022). Innovations in technology introduce a degree of uncertainty regarding their expected outcomes in the minds of potential adopters. However, they also represent an opportunity to reduce uncertainty through the informational underpinning of the technology.

2. Communication channel

These are the pathways through which information moves from one person to another. Communication plays a crucial role in the diffusion process, and communication channels refer to the means through which messages are transmitted among individuals. Although mass media channels are recognized for their rapidity and effectiveness, interpersonal channels are the most influential in persuading an individual to adopt a new idea (Jan, et al., 2011). It's worth noting that individuals tend not to assess innovations solely based on scientific studies but rather rely on subjective evaluations provided by other individuals. Effective communication typically occurs among homophilous individuals, those who share similarities. However, the person driving the adoption of the innovation, often referred to as the change agent, tends to be heterophilous, meaning they are different from the individuals they are persuading. This divergence may stem from their technical expertise or, at times, their educational and social status, setting them apart from the adopters (Srivastava & Moreland, 2012).

3. Time

The third aspect within the diffusion process is time, and time assumes a critical role in this process, covering three essential dimensions:

(i) The process of innovation decision-making,

(ii) The degree of innovativeness, and

(iii) The pace at which innovations are adopted.

Time isn't an isolated entity but rather an intrinsic aspect intertwined with events. It serves as an unmistakable element within any activity where the factor of time holds significance

4. Social structure

A social system can be described as a collection of interconnected entities working collaboratively to address shared challenges in pursuit of a common objective. The social structure encompasses various components such as individuals, groups, organizations, and subsystems. Within this framework, all members collaborate, at least to some extent, with the common aim of addressing shared problems and achieving mutual objectives. This shared purpose acts as the cohesive force that binds the system together. Importantly, diffusion unfolds within the context of a social structure. It serves as a boundary within which innovation spreads, defining the parameters of its reach and influence.

Moreover, the social system comprises key elements like opinion leaders and change agents. Opinion leaders wield significant influence within the system. Opinion leadership refers to an individual's ability to influence the attitudes or actions of others. Conversely, change agents have a role in convincing individuals to embrace innovations and also act as a barrier against the spread of unfavourable innovations.

Adoption process

Adoption is essentially the choice to fully utilize an innovation as the most favourable course of action (Ray, 2017). Adoption primarily involves a decision-making process. In rural villages, it's observable that even farmers who are aware of innovative practices don't immediately implement them; they take time to deliberate and act. This time lapse signifies a gap between knowledge and action when it comes to adopting new farming practices.

The decision to embrace an innovation seldom happens as a single, immediate occurrence. Instead, it evolves through a decision-making process involving several mental stages. Initially, the farmer becomes aware of the idea, and then proceeds to gather pertinent information and develop an attitude toward the innovation. Eventually, after these stages, a final decision is made, either to embrace or reject the innovation. It's worth noting that sometimes, even after initially deciding to adopt, a farmer may reverse their decision due to conflicting messages or information. Throughout all these stages, communication sources play an active role in assisting the farmer in making informed decisions.

Stages in adoption process

In the pioneering research on the spread of hybrid corn seed in two Towa communities in the United States, scientists brought attention to a sequential series of stages in the adoption process by farmers (Rogers, et al., 2019). These stages include:

Awareness: Initially, farmers gain knowledge about the existence of an innovation.

Conviction: Subsequently, they form a belief in the value and utility of the innovation.

Acceptance: This stage signifies a willingness to try out the innovation.

Trial: Farmers experiment with the innovation on a limited basis.

Adoption: Ultimately, the innovation is fully embraced and incorporated into their practices.

Initially, the first two stages were expanded into three, resulting in a five-stage model. This model, formulated by the Sub-Committee on Rural Sociology in the North Central region in the year 1955, gained widespread acceptance and garnered global attention as it effectively encapsulated the complex dynamics involved in the diffusion of farm practices. The adoption process consists of five stages: (i) awareness, (ii) interest, (iii) evaluation, (iv) trial, and (v) adoption. It is a process that occurs over a period of time and consists of a series of actions.

i. Awareness stage

A farmer may become aware of a new idea but lacks comprehensive information about it (Beal & Bohlen, 1957). While aware of the concept, the farmer may not possess a detailed understanding of what it entails, its benefits, and its functioning. For example, a farmer might have heard about SRI cultivation in rice but may not have a complete grasp of what SRI is, its purpose, and its operational mechanism.

ii. Interest stage

During the interest stage, a farmer's curiosity about the innovation intensifies, prompting them to actively seek more information (Beal & Bohlen, 1957). This additional information may be obtained from various sources, including extension officers, fellow farmers, or any other trusted source. At this point, the farmer aims to gain a deeper understanding of the innovation, including its nature, functionality, and potential benefits.

iii.Evaluation stage

During this stage, the farmer mentally assesses the relevance of the new idea to their current and anticipated future circumstances, ultimately making a decision on whether to experiment with it or not (Beal & Bohlen, 1957). The farmer evaluates the practicality and usefulness of the innovation. For example, after learning about SRI cultivation in rice and gaining insights into its components and how they enhance yield and water conservation, the farmer analyzes whether adopting SRI cultivation would indeed lead to improved rice yields.

iv. Trial stage

Farmers often refrain from immediately embracing a new idea or innovation on a large scale due to their aversion to risk, even when the idea shows promise. Instead, they opt for a cautious approach by initially experimenting with the new idea on a small scale to assess its practicality, suitability, and relevance to their specific circumstances. Even though farmers decide to give the idea a try based on its perceived benefits, they typically conduct small-scale trials in their own fields to effectively evaluate its effectiveness, even after extensive contemplation and information gathering.

v. Adoption stage

Upon being content with the performance of the new idea tested on a small scale in their own context, farmers proceed to integrate the new idea into their regular farming practices (Beal & Bohlen, 1957). This trial phase essentially serves as a practical assessment of the innovation. As the innovation proves advantageous, the farmer makes a definitive decision and incorporates it into their farming operations at a scale that aligns with their specific circumstances, maintaining its continued use as a standard part of their farming activities.

Innovation-decision process

The innovation-decision process characterizes the path taken by an individual (or another decision-making unit) as they progress from the initial awareness of an innovation. This journey includes the development of an attitude toward the innovation, reaching a final decision to either embrace or decline it, putting the new idea into practice, and ultimately reaffirming the decisions (Rogers, 2003).

The process of adopting or rejecting an innovation ultimately boils down to an individual's decision. When an individual chooses to adopt, they start incorporating a new concept, practice, or object into their routine and discontinue the use of the prior idea that the innovation supersedes.

The decision-making process associated with innovation adoption possesses unique characteristics that set it apart from other types of situations involving decision-making. When considering the adoption of an innovation, an individual is faced with the task of selecting from several available options, with the innovation representing a new option that didn't previously exist. This novelty and introduction of a new alternative distinguish the decision-making related to innovations from other decision-making situations.

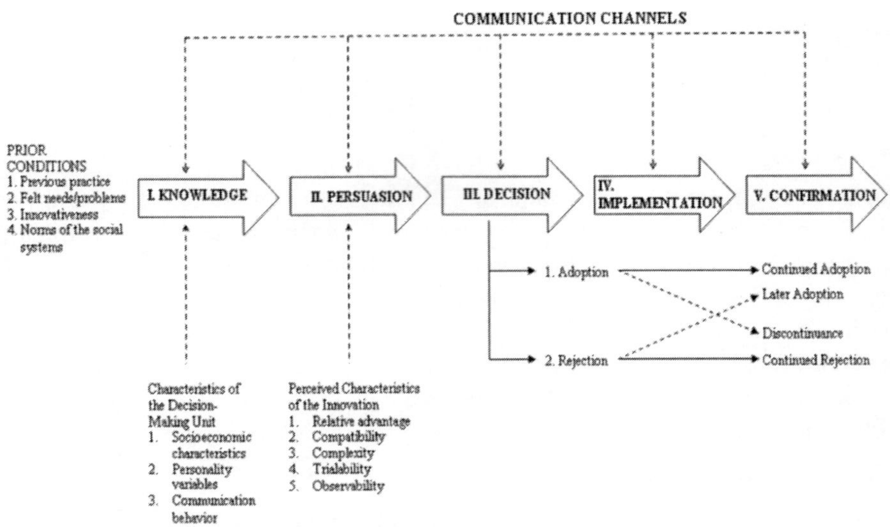

Framework outlining the phases in the innovation-decision process

1. Knowledge stage

Knowledge emerges when an individual (or a decision-making unit) encounters the innovation and acquires a certain level of comprehension regarding its operation (Rogers, 2003).

There are three distinct types of knowledge held by an individual that have an impact on their decisions:

i) **Knowledge of awareness** serves as a catalyst, motivating individuals to seek out information on "how-to" aspects and underlying principles. This phase of seeking information is primarily focused on the knowledge stage of the innovation-decision process, but it can also extend into the persuasion and decision stages.

ii) **Practical knowledge** encompasses the essential facts required for the proper utilization of an innovation. If individuals fail to acquire a sufficient level of how-to knowledge before experimenting with and adopting an innovation, it often leads to rejection or discontinuation. Change agents can play a distinctive role by highlighting the significance of practical knowledge during the trial and decision stages of the process.

iii) **Principle knowledge** encompasses facts related to the fundamental principles governing the functioning of the innovation. Acquiring principles knowledge not only enhances the long-term competence of individuals in assessing future innovations but also reduces the chances of misuse and discontinuation.

2. Persuasion stage

Persuasion occurs when an individual forms either a favourable or unfavourable attitude toward the innovation (Rogers, 2003). While the knowledge stage primarily involves cognitive processes (pertaining to knowing and understanding), the primary mode of thinking during the persuasion phase is affective, involving emotions and feelings. At this point, a comprehensive understanding of the innovation starts to form. The individual becomes emotionally invested in the innovation, which motivates them to actively seek information about this novel concept.

3. Decision stage

Decision-making occurs when an individual (or another decision-making entity) engages in a series of actions that lead to a final choice: whether to embrace or reject the innovation (Rogers, 2003). Adoption represents the choice to fully integrate the innovation as the most suitable course of action, while rejection entails the decision not to adopt the innovations.

In numerous instances, a small-scale trial plays a pivotal role in the adoption decision. This trial phase serves as a vital method to reduce the perceived uncertainty associated with the innovation for the adopter.

4. Implementation stage

Implementation occurs when an individual (or another decision-making entity) translates the innovation into practical use (Rogers, 2003). Until this stage, the innovation-decision process has largely been a cognitive process, but implementation represents the shift into tangible behaviour change as the new idea is actively put into practice.

5. Confirmation stage

Confirmation occurs when an individual seeks affirmation for a prior innovation-related decision (Rogers, 2003). However, this decision may be revisited if the individual encounters conflicting messages about the innovation. The confirmation stage is an ongoing process that continues beyond the initial decision to adopt or reject, lasting indefinitely.

In this stage, change agents take on the added responsibility of providing encouraging messages to individuals who have already adopted the innovation. Throughout the confirmation process, individuals aim to avoid or reduce internal conflicts or cognitive dissonance by adjusting their knowledge, attitudes, or behaviours.

Dissonance

Human behaviour change is, to a significant extent, motivated by a state of internal imbalance, a condition of mental discomfort that individuals instinctively seek to reduce or resolve (Ray, 2017). When confronted with such dissonance, individuals are typically motivated to alleviate it by making adjustments in their knowledge, attitudes, or actions. This phenomenon is particularly relevant in the context of innovative behaviour, where individuals may find themselves motivated to make changes in response to this internal imbalance. This may happen:

1. When an individual becomes aware of an existing need or issue and starts seeking information about possible solutions, like an innovation.

2. When an individual becomes aware of a new idea that aligns with their beliefs and values, they are motivated to adopt the innovation to resolve the dissonance between their beliefs and their current actions.

3. After deciding to adopt an innovation, an individual may seek further information that leads them to reconsider their initial choice.

These three approaches to reduce dissonance involve aligning attitudes and actions through behavioural adjustments. However, altering one's initial decision to adopt or reject can be challenging, as certain actions have already been initiated that tend to solidify the original choice.

Adopter categories

In a social system, individuals do not adopt an innovation at the same time. Instead, adoption occurs in a gradual sequence over time. To streamline the process of describing when individuals began using a new idea, adopter categories are employed. These categories classify members of the system according to their degree of innovativeness, grouping individuals with similar levels of innovativeness within each category (Ray, 2017).

A conventional method for classifying adopters and delineating the attributes of various adopter categories is presented here. Following that, the concept of innovativeness is examined as the underlying principle for categorizing adopters, enabling the classification of adopters and the depiction of their unique traits. The central criterion for adopter categorization revolves around innovativeness, which gauges how much earlier an individual or another adopting unit embraces new ideas compared to other members of the social system. When represented over time in terms of frequency, the adoption of an innovation typically conforms to a normal bell-shaped curve.

The allocation of adopters can be organized into five specific adopter categories by utilizing the mean and standard deviation as anchor points.

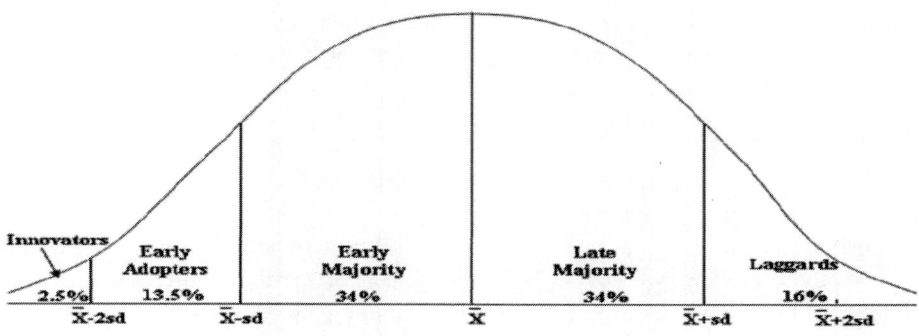

Categorizing Adopters Based on Their Level of Innovativeness

1. Innovators (venturesome)

These individuals constitute the initial 2.5% of the population in the system to adopt an innovation. Innovators exhibit a near obsession with venturesomeness. Their keen interest in novel concepts propels them beyond the confines of local peer networks and into more diverse social relationships that transcend geographical boundaries. It's common for innovators to establish communication patterns and friendships within a network of like-minded individuals, even if they are geographically distant (Ray, 2017).

Becoming an innovator requires several prerequisites. Having access to substantial financial resources proves beneficial, as it helps absorb potential losses from unprofitable innovations. Additionally, an innovator must possess the ability to grasp and apply complex technical knowledge. Consequently, innovators serve as gatekeepers, facilitating the flow of fresh concepts into the system and thus contributing significantly to the process of innovation adoption.

2. Early adopters (respectful)

This group encompasses the subsequent 13.5%, falling within the range between the mean minus one standard deviation and the mean minus two standard deviations. Early adopters are more deeply integrated into the local social system compared to innovators, who tend to have a broader, cosmopolitan outlook. Early adopters, on the other hand, are deeply rooted in their local communities. Among all adopter categories, early adopters typically exhibit the highest level of adoption leadership within most systems. Many individuals regard the early adopter as the go-to person to consult before embracing a new idea (Ray, 2017).

Change agents actively seek out early adopters as local champions to expedite the diffusion process. Since early adopters are not significantly ahead of the average individual in terms of innovativeness, they serve as role models for many other members of the social system. Early adopters enjoy respect from their peers and are seen as exemplifying successful and discreet utilization of new ideas. Furthermore, early adopters reduce uncertainty about new ideas by adopting them and then sharing their subjective evaluations of the innovation with close peers through interpersonal networks.

3. Early majority (deliberate)

The early majority consists of individuals who adopt new ideas just before the average member of a system does. Comprising 34% of the adopters, the early majority falls within the range between the mean time of adoption and mean minus one standard deviation. They engage in frequent interactions with their peers but rarely hold positions of opinion leadership within the system. Instead, they tend to follow the lead of others (Ray, 2017).

These individuals have relatively limited contact with cosmopolitan sources of information. These neighbours and friends often serve as the primary sources of information about innovations for the majority of community members.

4. Late majority (skeptical)

The late majority comprises individuals who adopt new ideas just after the average member of a system does. Similar to the early majority, the late majority accounts for roughly one-third of the system's population. The subsequent 34% of adopters are categorized as the late majority. They occupy the space between the mean and one standard deviation behind the mean. Their adoption of innovations may be driven by economic necessity and increasing peer pressure (Ray, 2017).

Late majority individuals tend to approach innovations with skepticism and caution, and they are not quick to adopt until the majority of others in their system have already done so. Given their relatively limited resources, the late majority requires most of the uncertainty surrounding a new idea to be resolved before they consider it safe to adopt.

5. Laggards (traditional)

The remaining 16% of adopters, positioned to the right of the mean plus one standard deviation, are commonly known as laggards. They represent the final segment of a social system to adopt an innovation and typically exert very little influence on others' opinions. They are the most deeply rooted in their local environment among all adopter categories and often have limited social connections within their system. Laggards tend to be resistant to change, with their frame of reference firmly rooted in the past (Ray, 2017).

It is important to note that the term "laggard" should not be interpreted as a derogatory label. Most individuals outside of the laggard category possess a strong bias in favour of innovation. Scholars who study diffusion and employ adopter categories do not intend to disrespect laggards. Instead, they recognize that these individuals face unique challenges in adopting innovations, and the term "system blame" may better capture the reality of their situation.

Conclusion

In conclusion, having a deep understanding of the adoption and diffusion processes is instrumental for extension agents in their efforts to facilitate and expedite these processes. Diffusion is a unique form of communication that deals with disseminating messages perceived as novel ideas. The diffusion and adoption of innovations represent intricate processes influenced by various factors, including the inherent characteristics of the innovation, the traits of the adopters, and the wider social and organizational context. Consequently, there's often a level of uncertainty and perceived risk associated with the diffusion process. Individuals can mitigate this uncertainty by actively seeking information. Adoption, on the other hand, is the decision to fully embrace an innovation as the most suitable course of action available. Essentially, adoption is a decision-making process in itself. By comprehending the intricacies of adoption and diffusion, extension agents are better equipped to expedite the adoption of innovations in the field of agriculture.

References

Beal, G. M., & Bohlen, J. M. (1957). The diffusion process. Ames: Iowa State University of Science and Technology. 111-121. https://core.ac.uk/download/pdf/7044374.pdf

Jan M., Sultan K., Ali S. (2011). Role of Communication in Diffusion and Adoption of Agricultural Innovations. Gomal University Journal of Research, 27(1): 111-118. http://www.gujr.com.pk/index.php/GUJR/article/view/446

Ray, G. L. (2017). Adoption and Diffusion of Innovations. Extension Communication and Management. (8th ed.). 157-182. Kalyani Publishers.

Rogers, E. M. (2003). Diffusion of innovations (5th ed.). New York: Free Press.

Rogers, E., Singhal, A., & Quinlan, M. (2019). Diffusion of Innovations. 182-186.

Rohilla, S. Mukteshawar, R. & Mishra, A. (2022). Diffusion and Adoption of Innovations. 10.22271/ed.book.1831.

Srivastava J., & Moreland J. J. (2012). Diffusion of Innovations: Communication Evolution and Influences. The Communication Review. 15(4). 294-312. https://doi.org/10.1080/10714421.2012.728420

8

Mobile Applications: Transforming Extension Services in India

Rishabh Singh Gaur[1]*, Biswajit Mallick[2], Rubina Rai[3] andAmimsha Limboo Subba[4]

[1]PhD Scholar, Department of Agricultural Extension Education, ICAR-NDRI, Karnal, Haryana, 132001
[2]PhD Scholar, Department of Agricultural Extension Education, College of Agriculture, Odisha University of Agriculture and Technology, Bhubaneswar, Odisha, India
[3]Young Professional – 1, ICAR - NRC for Orchids, Pakyong, Sikkim – 737106
[4]Research Scholar, Department of Extension Education, Dr. Rajendra Prasad Central Agricultural University Pusa, Samastipur- 848125, Bihar, India.
**Corresponding author's email:* rishabhsg98@gmail.com

Abstract

The agriculture sector in India is struggling with numerous obstacles, such as less production compared to high demand for food, reduced agricultural labour force, climate change, and disease and pest infestation. To overcome such difficulties, concern should be given to efficient and sustainable production techniques, advanced agricultural practices, and the involvement of artificial intelligence in agriculture. Agriculture extension plays a vital role in providing the correct information about agricultural practices along with procurement and marketing strategies which profitable for farmers. Public, private and non-government organizations are working towards developing the farming community. For that, various approaches are there to organize agricultural extension services. But recently, the usability of basic mobile phones, smartphones, and multiple applications is producing significant results in accurately and timely disseminating information to needy consumers. Mobile application-based extension services are gaining higher importance among the country's farming community. Cheaper cost, easy availability, and accessible information facilities are crucial for the success of mobile application-based extension services. This chapter highlights public and private sector-based mobile advisory services, their advantages and constraints to Indian agriculture.

Keywords: *Agriculture, Extension Services, Mobile Advisory Services, Mobile Application.*

Introduction

In modern agriculture, soft resources like knowledge and skills are sometimes even more crucial than hard resources like inputs. In the 21st century, agriculture has to overcome a number of obstacles, including feeding an expanding population while a declining rural labour force, producing more feedstocks for a huge bioenergy market, supporting the overall development of many developing countries heavily dependent on agriculture, adopting more efficient and sustainable production techniques, and adapting to climate change (Colle, 2011). In order to make efficient and effective agricultural decisions, farmers are increasingly adopting information (Cash, 2001). Compared to the current availability of roughly 0.1 million personnel, the required number of extension agent who worked in field level is anticipated in between 1.3 to 1.5 million. Mobile application-based extension services provide a practical answer to these new problems in global agriculture (Donner and Jonathan, 2008). In this chapter, we will discuss the various mobile applications facilitating extension services for the betterment of agriculture.

Overview of Extension Services

Extension services facilitate technical assistance for the farmers in agricultural sector. It aims to give people access to the crucial inputs and services that help and boost their output. Agricultural extension efforts spread innovative concepts and methods created by agricultural scientific research centres among farmers. To make their crops less vulnerable to pests and diseases, smallholder farmers worldwide require access to agrarian guidance (Meena and Singh, 2013). Employees who provide consulting and extension services are crucial in informing communities about science-based solutions (Hanumankar and Hemnath, 2011). Their advice on crop health and identifying pests and diseases aids farmers in losing less of what they produce, boosting agricultural output.

Types of Extension Services

In India, extension services fall into three primary areas.

1. **Technology transfer services:** Charged with providing farmers guidance, information, and knowledge.
2. **Advisory Services:** In charge of aiding farmers in the event that they encounter specific difficulties.

3. **Facilitation Services:** Assisting farmers with problem-solving by helping them analyze their issues.

Extension Services in India: The following organizations offer Extension Services in India:

- Public sector organizations like Ministry of Agriculture & Farmers Welfare or Ministry Rural Development;
- Private organizations like input dealers (seed and fertilizer), agro-based advisors, the crop researchers, technology providers;
- Non-profit or non-government organizations, commodity based boards (Coffee Board, Rubber Board), or farmer-producer organizations.

Public Extension Services: Central and State government organizations are responsible for providing the extension services in public extension service system. Also, ICAR institutes, Krishi vigyan kendras, state agriculture universities, state agriculture departments etc. are disseminate their available services to farmers.

Private Extension Services: Private Extension Services include advice, credit, insurance, contract farming, pesticides, seeds, fertilizers, and machinery. Additionally, a number of private businesses, including cooperatives, progressive farmers, and farmers' organizations (CIGs, FPOs), are direct participants in this system. Other private players include agribusiness owners, input distributors, nongovernmental organizations (NGOs), private banks, private media, donor organizations, consulting firms, etc.

Different Strategies for Organizing Services of Agricultural Extension

Various methods can be used to organize agricultural extension services. Factors such as administration framework, agricultural research, and training, goals and objectives of extension, grassroots level infrastructure, and availability of funding and human resources have influenced the extension service. The following eight models or techniques for organizing agricultural extension have been recognized by Axinn (1988):

1. **The general strategy for agricultural extension:** Most nations follow this standard paradigm. It adopts a top-down approach to technology transfer with the presumption that informing farmers about new technology accessible through research institutes will lead to higher productivity and production.
2. **Commodity-specific strategy:** These strategies concentrate about a single product, such as rubber, coffee, and sugarcane, etc. The strategy is highly specialized in nature, with the commodity organization in charge of the program's development and execution.
3. **The training and visit strategy** gained popularity in the middle of the

1970s. This strategy strongly emphasises extension agents visiting specific contact farms and receiving ongoing training in the most recent methods. The planning and execution of the program are centralised.

4. **Participatory extension strategy:** It places a strong focus on creating farmers' organizations and groups, such as self-help groups, as well as absorbing farmers' knowledge. Farmers participate in local program control and decision-making connected to its design and implementation.
5. **Project strategy:** In this case, more outside resources are infused while extension efforts are focused on a specific location for a predetermined time. The project staff is in charge of designing and carrying out the program.
6. **Agro-Ecosystem strategy:** This strategy's primary goal is to design appropriate technologies for farmers in resource-limited places collaboratively. The technologies are created using a comprehensive agro-ecosystem that includes researchers, extension specialists, and farmers of all genders.
7. **The cost-sharing strategy:** Farmers' expenses for agricultural extension are covered by the local population and the implementing agency.

Current Extension Service System

State Departments of Agriculture are primarily in charge of providing agricultural extension services in our nation. Input-supplying companies, finance institutions, and nonprofit organizations have all started to offer agricultural extension services in recent years. State departments of agriculture were in charge of extension, agricultural education, and research up to the 1960s. Agriculture-related research and instruction were then moved to agricultural universities. Additionally, the Department of Agriculture & Farmers Welfare has been split up into Departments/ Directorates, like the Department of Horticulture, Department of Agricultural Engineering, Directorate of Oilseeds and Pulses, etc., due to increased specialization and workload.

Perforation of Smart Phones in Agriculture

Smartphones' usefulness, usability, and affordability have helped them become one of the most popular new technologies in recent years among various user sectors. In addition to making phone calls, a smartphone can run custom software, send and receive e-mail, connect to Wi-Fi and the internet, view Office documents, and access the internet. One other crucial aspect of a smartphone is its "user interface." Increasing at a compound annual growth rate (CAGR), it is predicted that India's population of smartphone users will more than double to 859 million by 2022 from 468 million users in 2017 of 12.9% (ASSOCHAM-PwC, 2019). With 560 million internet subscribers in 2018, India is the second largest internet user behind China (IAMAI, 2019). According to the Ericsson Mobility Report, 1.4 billion mobile subscribers are anticipated to be in place by 2021 (CNBC, 2016). Over the past ten years, mobile phones, tablets, and other omnipresent technologies have inundated marketplaces in developed and developing nations, including India (TRAI, 2015). The technologies connected the entire farming society with available 2G and 3G networks (Vodafone, 2010). These mobile-based smart applications have the potential to provide various subscribers, including farmers, traders, and producers, with timely information. Weather, rainfall, and crop statistics are all included in the information supplied, and some programs also assist in updating market data for commodity prices and facilitating local buying and selling via hand-held devices.

Mobile Applications are a boon for Extension Services!

Mobile applications/mobile apps are programmed software used to operate smartphones and tablets but are identical to those found on desktop and laptop computers (Costopoulou et al., 2016). They carry out crucial and particular tasks, such as enhancing productivity, providing amusement, and facilitating information access. They offer consumers mobile content, including text, audio, recordings, photos, graphics, and videos, and are created to be interactive. Numerous mobile applications for agriculture have been made. Mobile phones can be accessible in rural areas also (BCG, 2016).

Sennuga (2019) highlighted that mobile application are prominently used for linking extension personnel with rural farmers due to inefficiency and ineffectiveness of the traditional extension systems (T&V approach). In order to increase agricultural output and speed up development programs in rural regions, mobile phones have emerged as one of the ICTs with the quickest growth rates (Asa and Uwem, 2017). Ogunniyi and Ojebuyi (2016) recognized that mobile phone use in agricultural extension can aid in the eradication of poverty and promote socioeconomic advancement.

Smallholders can use mobile apps to access current, real-time information, either directly or through advisory services. By giving them access to pertinent information, digital development, such as agricultural mobile apps can assist advisory service professionals in offering farmers the best advice possible (Torero & Braun, 2006). Utilizing mobile phone applications is a successful strategy for quickly reaching clients (farmers). Mobile phone applications can reach a large audience of farmers without having to make expensive visits to their farms, making them a valuable addition to the T&V extension strategy.

Equally, agricultural actors have created group chats on social media platforms like WhatsApp, Facebook, Telegram, and Instagram to share information and market agricultural products (Kashem, 2011). These mobile applications allow extension groups to use these platforms to run extension programs. With the help of mobile phone applications, extension personnel can quickly reach out to larger farmers with pertinent information.

Overview of Mobile Applications for Agricultural Extension

1. Public Sector Mobile Advisory Services

1.1. Farmer Call Center (Kissan Call Center)

The Department of Agriculture & Cooperation (DoA&C), Ministry of Agriculture, Government of India, opened Farmer Call Centers across the nation on January 21, 2004. These call centres serve to address concerns voiced by farmers in the local tongue immediately. Every state has call centres handling calls from all across the nation. These call centres handle inquiries regarding agriculture and associated industries. Farmers can contact with toll free number 18001801551 and get information within 24 hours .

1.2. Price of the Market by SMS from Rubber Board, India

Through SMS, the Rubber Board updates domestic and foreign natural rubber prices for the nation. Through this service, rubber producers and traders in India (particularly those in the southern state of Kerala) can receive real-time SMS alerts about price changes.

1.3. Virtual Krishi Vigyan Kendra, or vKVK

Agro-advisories can be provided to farmers' mobile phones utilizing SMS alerts and voice-based advice using the vKVK (Virtual Krishi Vigyan Kendra) platform based on a basic messaging system. The vKVK platform uses the internet and mobile technology to link KVKs with farmers. A voice-based notice or recommendation can be delivered to farmers via phone call thanks to a phone-based delivery system. Under the direction of KVK experts, a recorded message may be sent to all farmers via the vKVK platform. It is tested in some districts of Uttar Pradesh, Uttarakhand and Karnataka (ICRISAT, 2012). 1000 beneficiaries along with farmers, extension personnel, and officials of department from Gwalior enrolled in Kisan Mobile Sandesh (KVK, RVSKVV, 2012).

1.4. mkisan

The mkisan initiative challenge grant has helped to launch the mKisan project. The mKisan project is being carried out by the International Livestock Research Institute (ILRI), India, in collaboration with Handygo technologies, a mobile value-added service provider, CABI South Asia, and Digital Green, an NGO for video-enabled extension (ILRI, 2012). The goal of the project is to create comprehensive agro-advisory services for Indian smallholders to have access to a mobile phone (IANS, 2008). The project aims to deliver daily bulletins on agro-meteorology, crop pest and livestock disease outbreaks, market information, information on local service-providing sources, and information access to women farmers.

1.5. Kisan Help Line

The BAU (Bihar Agriculture University), Sabor, has operated an assistance line for local farmers since 2012. Farmers can call the agricultural experts from several streams at the helpline number 0641-2451035 with their questions. The helpline is open every working day from 10 a.m. to 5 p.m.

2. Private Sector Mobile Advisory Services

2.1 Lifelines India

2,00,000 farmers in 2130 villages in four Indian states are connected via an inventive combination of the internet and telephony, according to the website lifelines-india.net, which also has eight crop cycles that have been completed and more than 4,00,000 questions and answers in its knowledge base. It collaborates with NGOs and other organizations to operate. Lifelines India- Soochna Se Samadhan (Solutions via knowledge) aims to leverage the power of speech as the main channel for distributing knowledge.

2.2 Fisher Friend

With the help of the M. S. Swaminathan Research Foundation (MSSRF), Tata Teleservices (Parashar, 2009), Asute System Technology, and QUALCOMM, coastal fishing communities now have access to mobile-based advisory services that provide immediate access to useful information like weather conditions, where they can and cannot fish, and market price information.

2.2 IKSL, or IFFCO Kisan Sanchar Limited

IKSL via means of voicemails delivered in the regional tongue. 95 000 voicemails have been sent, 81 000 questions and answers have been posted, with 5,000 farmer feedback notes. IKSL has almost four million members, with 40,000 cooperative societies functioning as IKSL Retailers. 1.3 million active farmers are benefiting from IKSL's value-added services (Tiwari, 2012).

2.3 mKRISHI "a product of TATA Consultancy Services"

Three services are offered by mKrishi to farmers: an Interactive Voice Response System (IVRS), mobile technology, and automated weather station information integration service (The Financial Express, 2013).

2.4 Nokia Life Tools

For consumers in emerging markets, small towns and rural areas, Nokia Life Tools is a collection of services that includes agriculture, education, and entertainment services. The service delivers timely and pertinent information directly to users' mobile phones that is tailored to their location and personal preferences.

2.5. Knowledge Help Extension Technology Initiative (KHETI)

The Sironj Crop Producers' Company (SCPCL), a cooperative of small farmers from the villages surrounding Sironj in the Madhya Pradesh State of India, has been implementing KHETI since August 2008 with financial assistance from the UK Engineering and Physical Science Research Council. KHETI offered solutions by combining desktop computers, the internet, mobile devices, and Interactive Voice Response System (IVRS) (Dearden et al., 2012).

Advantages of Mobile Applications in Extension Services

Affordableness, widespread ownership, voice communication, and quick\ convenient service delivery are some of the benefits of mobile apps. Due to these factors, there is an explosion in the number of mobile apps worldwide, made possible by the development of mobile networks, the expansion of mobile device functionality, and the decline in the price of mobile handsets (World Bank, 2012). Farmers always have access to a wealth of information about crops, soil, weather, rainfall, seeds, and equipment at their fingertips. Software solutions can help farmers and their advisors run their farms more effectively by swiftly collecting data, evaluating it, and producing a set of recommendations based on the analysis (Patel and Patel, 2016). Farmers cannot access the material, which has been collated and well arranged. The information is prepared and extremely well organized so that the farmer doesn't have to spend time looking up and referring to it (Rao, 2007). The visibility and awareness of prospective buyers and sellers in the area and the chance to establish personal relationships promote market connection. A real-time form of delivery is available for commodity prices.

By keeping track of dead stock, updating their inventory, and taking notice of purchase requirements, mobility can help farmers create better warehouse facilities, allowing them to meet delivery promises on time, get their goods to the final consumer, and maintain product quality (Saravanan and Bhattacharjee, 2014). Cell phones have an even more significant effect on price dispersion for participants farther from their markets and those with less favourable transportation conditions (TNAU, 2013). Additionally, farmers can monitor orders and be well-informed about their investments.

Drawbacks of Mobile Applications in Extension Services

The lack of relative advantages, compatibility, trialability, observability, and complexity of mobile apps make it tough and detrimental for users to access the programs without difficulties (Veeraraghavan, et al., 2009). Due to the diversity of languages, translation will be necessary at every stage, even if the best applications do not support regional languages. This will create dependency, which will lower acceptance and popularity. It is occasionally possible for network difficulties, data transmission speed, and legal restrictions to prohibit farmers from receiving the most recent and comprehensive information. A trained individual may be needed to comprehend and translate complex tasks that must be carried out on a farm, unclear material, and movies in other languages.

Conclusion

Access to information is changing due to farmers' increasing use of mobile applications. Farmers have been captivated to acquire new apps to stay up with contemporary technology since the digital revolution and internet accessibility in rural areas. Various innovative apps are emerging in response to evolving needs and difficulties in the agricultural and related sectors. It is crucial to be picky when selecting an app, check to see if it delivers reliable and up-to-date information, and ensure it complies with criteria as the number of apps keeps rising. Due to the context-based nature of the agricultural activity, which is essentially identifiable by different geographic locations, smartphone applications now available in one context can be created to fit other crops, nations, or regions. As a result, mobile apps should focus on comprehensive rural development, create stronger bonds between farmers and consumers through technology that takes gender equality into account, and teach and equip farmers through technology-driven platforms for activities that generate money.

References

Asa, U. A., & Uwem, C. A. (2017). Utilization of mobile phones for agricultural purposes by farmers in Itu Area, Nigeria. European Scientific Journal, 13(9): 395-402. http://dx.doi.org/10.19044/esj.2017.v13n19p395

ASSOCHAM-PwC. (2019). Emerging technologies disrupting the financial sector. Background Paper. May, 2019. Conference on Communication Systems and Network Technologies (CSNT), Rajkot, India. https://www.pwc.in/assets/pdfs/consulting/financial-services/fintech/publications/emerging-technologies-disrupting-the-financial-sector.pdf

Axinn, G.H. (1988): Guide on Alternative Extension Approaches, FAO, Rome.

BCG, (2016). The rising connected consumer in rural India by Nimisha Jain and Kanika Sanghi. https://www.bcg.com/publications/2016/globalization-customer-insight-rising-connected-consumer-in-rural-india

Cash, D. W. (2001). In Order to Aid in Diffusing Useful and Practical Information: Agricultural Extension and Boundary Organizations. Sci. Tech. Hum. Values, 26(4): 431-453. Doi: https://doi.org/10.1177/016224390102600403

CNBC, (2016). How India is shaping the global smartphone market by Harriet Taylor 21 September 2016. https://www.cnbc.com/2016/09/21/how-india-is-shaping-the-globalsmartphone-market.html

Colle, D. R. (2011). Book Review on ICTs for Agricultural Extension: Global Experiments, Innovations and Experiences, (Ed) R. Saravanan (2010). Journal of Development Communication, 22(1). https://saravananraj.in/2020/03/icts-for-agricultural-extension-global-experiments-innovations-and-experiences/

Costopoulou, C., Ntaliani, M. & Karetsos, S. (2016). Studying Mobile Apps for Agriculture Informatics Laboratory, Department of Agricultural Economics and Development, Agricultural University of Athens, Greece. IOSR Journal of Mobile Computing & Application (IOSR-JMCA) e-ISSN: 2394-0050, P-ISSN: 2394-0042, 3(6): 44-49.

Donner and Jonathan. (2008). Research Approaches to Mobile Use in the Developing World: A Review of the Literature. J. Inform. Soc. 24(3):140-159. Doi: https://doi.org/10.1080/01972240802019970

Hanumankar and Hemnath R. (2011). Application of ICT in Agricultural Extension: An evaluation study of Kissan Call Centres (KCCs).Information and communication technology for agriculture and rural development. Edt. Saravanan, R., Kathiresan, C., and Indra Devi, T., New India Publishing Agency, New Delhi, 198-206.

IAMAI, (2019). India Internet Report-2019. Internet and mobile association of India. https://www.iamai.in/KnowledgeCentre

IANS. (2008). Farmers to Get Market Rates on BSNL Mobiles.

ICRISAT, (2012). Agropedia: The Knowledge and Interaction Hub for Indian Agriculture. Information Bulletin on Agropedia.

ILRI Project File, (2012). mKisan-Using Mobile Technologies to Strengthen Farmer-Extension Expert-Linkages in India. http://cgspace.cgiar.org/bitstream/handle/10568/24461mkisan Mobile Oct2012.pdf

Kashem, M. A. (2011). Farmers' Use of Mobile Phones in Receiving Agricultural Information towards Agricultural Development.

KVK, RVSKVV, (2012). KVK's brief achievement 2009-'10 to 2011-12. http://www.rvskvv.nic.in/Documents/DES/achievement_09-12.pdf

Meena, H. R., & Singh, Y. P. (2013). Importance of information and communication technology tools among livestock farmers: A review. Scientific Journal of Pure and Applied Sciences, 2(2), 57-65. Doi: http://dx.doi.org/10.14196/sjpas.v2i2.466

Ogunniyi, M. D., & Ojebuyi, B. R. (2016). Mobile phone use for agribusiness by farmers in Southwest Nigeria. Journal of Agricultural Extension, 20(2): 172-187. https://doi.org/10.4314/jae.v20i2.13

Parashar, P. (2009). Tata Teleservices, Impetus join hands to launch Mandi Bhav for farmers. The Financial Express. Accessed on 8 july 2023. https://www.financialexpress.com/archive/tata-teleservices-impetus-join-hands-to-launch-mandi-bhav-for-farmers/414058/

Patel, H. and Patel, D. (2016). Survey of android apps for agriculture sector. International Journal of Information Sciences and Techniques, 6(1-2), 61-67.

Rao, N. H. (2007). A framework for implementing information and communication technologies in agricultural development in India. Technological Forecasting and Social Change, 74(4), 491-518. https://doi.org/10.1016/j.techfore.2006.02.002

Saravanan, R., & Bhattacharjee, S. (2014). Mobile phone applications for agricultural extension in India. Mobile phones for agricultural extension: Worldwide mAgri Innovations and Promise for Future. (Edited by Saravanan, R.), 1-75.

Sennuga, S. O. (2019). Use of ICT among Smallholder Farmers and Extension Workers and its Relevance to Sustainable Agricultural Practices in Nigeria. PhD Thesis. Coventry Universityin association with the Royal Agricultural University.

The Financial Express, (2013). Tata Teleservices Launches New Service 'Mandi Bhav'.

Tiwari, R. (2012). Government Ropes in IFFCO Kisan Sanchar Limited to Improve Kisan Call Centres. The Economic Times. Published on 5th May, 2012. https://economictimes.indiatimes.com/news/economy/agriculture/government-ropes-in-iffco-kisan-sanchar-limited-to-improve-kisan-call-centres/articleshow/13009174.cms?from=mdr

TNAU, (2013). Development of ICT based Tools/Technology towards an Interactive Multimedia Agriculture Advisory System. Accessed on 6th August, 2013. https://agritech.tnau.ac.in/govt_schemes_services/aas/index.html

Torero, M. & Braun, J. V. (2006). Information and communication technologies for development and poverty reduction – The potential of telecommunication. The Johns Hopkins University Press and IFPRI, Washington- DC.

TRAI. (2015). Highlights of telecom subscription data as on 31st August, 2015. Telecom Regulatory Authority of India (2015). Press Release No. 56/2015. http://www.trai.gov.in

Veeraraghavan, R., Yasodhar, N., & Toyama, K. (2009). Warana Unwired: Replacing PCs with Mobile Phones in a Rural Sugarcane Cooperative. Information Technologies and International Development, 5(1): 81-95. https://itidjournal.org/index.php/itid/article/download/327/327-802-2-PB.pdf

Vodafone, UN. (2010). Wireless Technology for Social Change: Trends in Mobile Use by NGOs.

World Bank. (2012). Mobile applications for agriculture and rural development. Washington, D.C. World Bank Group. https://documents1.worldbank.org/curated/en/167301467999716265/pdf/96226- REVISED-WP-PUBLIC-Box391469B-Mobile-Applications-for-ARD-v8S-Complete.pdf

9

Exploring Monitoring and Evaluation in Agricultural Extension

Jyotishree Anshuman[1*] and Shreya Kar[2]

[1]Ph.D. Scholar, Department of Agricultural Extension Education, Odisha University of Agriculture and Technology, Bhubaneswar, Odisha, India
[2]Msc. Scholar, Department of Extension Education, Punjab Agricultural University, Ludhiana, Punjab, India
[*]Corresponding author's email: 1997j.anshuman@gmail.com

Abstract

The book chapter titled "Exploring Monitoring and Evaluation in Agricultural Extension" provides a comprehensive exploration of essential concepts and practices within the domain of programme management. In the first section, the chapter delves into the fundamental definitions of monitoring and evaluation, establishing a strong foundation for readers to grasp the distinct yet interconnected nature of these two critical processes. It goes on to elucidate the different types of monitoring, encompassing process, impact, beneficiary, financial and administrative monitoring, while shedding light on their respective roles in gauging programme progress and effectiveness. Moreover, the chapter outlines purpose of monitoring and the step-by-step approach to effective monitoring, providing readers with practical guidance on data collection, indicator selection, and analysis. Moving into the realm of evaluation, the chapter examines the definition and rationale behind conducting evaluations of extension programmes. It delves into the reasons and timing for evaluation, emphasizing its role in accountability, learning, and continuous improvement. Additionally, it addresses the barriers that may impede successful evaluation and offers insights into overcoming them. The discussion on different types of evaluation provides readers with a comprehensive understanding of the various evaluation methodologies and their respective applications. The chapter culminates with an exploration of evaluation models, including the CIPP (Context, Input, Process, Product) model, the Kirkpatrick's Four-level Training Evaluation Model, etc., shedding light on their unique strengths and suitability for diverse evaluation contexts. The book chapter serves as a valuable resource for practitioners, policymakers, and researchers involved in extension programme management. It mentions the key differences between monitoring and evaluation and underscores their pivotal role in driving the success, sustainability, and impact of extension programmes.

Keywords: *Monitoring, Evaluation, Extension Programmes.*

Introduction

Extension programmes are strategically planned and implemented by an organization, typically a government department, with the majority of their funding originating from public sources. In the dynamic and ever-changing landscape of developmental initiatives, extension programmes play a pivotal role in fostering progress and transforming societies. These initiatives, designed to extend knowledge, skills, and resources to target communities, are instrumental in addressing pressing challenges and promoting positive change across various sectors. In order to justify the allocation of public funds and to measure effectiveness and sustainability of the programmes, it is imperative to conduct thorough and comprehensive monitoring and evaluations of their administration and impact at regular intervals.

Monitoring and Evaluation are two intertwined and complementary processes that form the backbone of effective programme management. Monitoring involves the continuous and systematic collection of data to assess the progress and implementation of the programme in real-time. It provides valuable insights into the immediate outcomes and helps in identifying any deviations from the planned activities. On the other hand, Evaluation is a rigorous assessment of the programme's overall impact, effectiveness, and efficiency, with a focus on understanding the achieved outcomes and long-term consequences. Together, these processes create a feedback loop, enabling stakeholders to make informed decisions, improve interventions, and maximize the positive impact on the target beneficiaries. Linking Monitoring and Evaluation is crucial because it bridges the gap between the planned intentions and actual outcomes. In this chapter, we delve into the essential aspects of Monitoring and Evaluation of Extension Programmes.

Definition of monitoring

Monitoring within social development can be described as the methodical and ongoing process of gathering and analyzing information about the advancement of a project or initiative over a period. (Gosling and Edwards, 2003) .This practice serves the purpose of ensuring that all relevant stakeholders are well-informed about the progress of the development intervention. Additionally, monitoring facilitates timely management decisions based on the available data and insights.

Different types of monitoring

Process or performance monitoring

Process monitoring in the context of development interventions revolves around observing and evaluating the activities undertaken as part of the intervention. Its primary objective is to determine whether these activities are being executed as planned and to what extent. Additionally, process monitoring encompasses the monitoring of resource utilization.It enables the assessment of the effectiveness of implementation, identification and resolution of issues and challenges, and seizing opportunities as they emerge. In essence, process monitoring serves as a valuable tool for ensuring the smooth and successful execution of projects and programs in social development.

Results or impact monitoring

Impact monitoring involves the continuous evaluation of the transformations generated by a project or program. Its primary focus is on assessing changes within the target population, which may include individuals, communities, supported organizations, or specific decision-makers. The purpose of impact monitoring is to gauge the progress made towards achieving the goals and objectives of the initiative, as well as to identify any unintended changes that may have occurred.

Beneficiary monitoring or beneficiary contact monitoring

It is a specialized form of impact monitoring focused on tracking the views and experiences of individuals or groups benefiting from a project or programme (IFRC 2011). This monitoring approach involves the use of beneficiary feedback mechanisms and beneficiary complaints mechanisms. Essentially, beneficiary monitoring falls under the broader umbrella of participatory monitoring and evaluation (M&E), emphasizing the active involvement of beneficiaries in assessing and providing feedback on the effectiveness and impact of the initiatives that directly affect them.

Financial monitoring

It pertains to the oversight and management of budgets and financial resources, often in connection with auditing processes. Its primary focus is on tracking and evaluating expenses relative to predefined categories of expenditure.

Management Information Systems (MIS)

This is typically linked to computer-based systems, although certain Civil Society Organizations (CSOs) may utilize paper-based versions. An MIS has the capacity to encompass various aspects of monitoring, but it is particularly prevalent in process, financial, administrative, and compliance monitoring.

Administrative or logistics monitoring

It involves overseeing various aspects related to the smooth functioning of an organization. This monitoring encompasses tasks such as maintaining premises, managing transportation, overseeing personnel, ensuring efficient stock-keeping, and handling other administrative functions.

Purpose of programme monitoring

Programme monitoring has diverse objectives, contingent on the monitoring type and particular circumstances. Nevertheless, its main aim is to consistently collect information about the advancement of a project or programme.This collected data is then used to make continuous adjustments that support effective management. This includes ensuring that plans and budgets are adhered to and making necessary modifications when required. In essence, effective monitoring is closely intertwined with competent management, as both are vital components for the successful implementation of a project or programme. By maintaining a robust monitoring process, organizations can stay agile and responsive, making informed decisions to optimize their initiatives and achieve desired outcomes.

Under various conditions, monitoring can fulfil a range of additional objectives. Some of the more prevalent ones include:

1. Keeping stakeholders informed of progress towards established goals and objectives to showcase accountability to diverse stakeholders.
2. Offering data to support adaptable decision-making in response to changing circumstances and conditions.
3. Assessing the necessity for adjustments in goals, plans, objectives or budgets over time.
4. Regularly testing assumptions of projects/programmes to ensure validity and effectiveness.
5. Identifying potential gaps in information, prompting the need for further research or data gathering.
6. Enhancing continuous learning both within the project or programme and beyond by providing valuable insights and feedback.
7. Demonstrating and communicating project activities and developments to external audiences for fundraising or marketing purposes.

Steps in Monitoring

1. Planning
2. Collecting information
3. Recording information
4. Storing information

5. Analyzing information
6. Sharing information

Evaluation

The term 'evaluation' finds its roots in the Latin word "valere," denoting strength or valor.

Definition of evaluation:

Michael Scriven (1991) described that Evaluation involves the methodical utilization of scientific approaches to analyze the program's design, implementation, enhancement, or results.

Daniel L. Stufflebeam (1973) defined that Evaluation is the procedure of defining, acquiring, and delivering valuable information to assess various decision options. Top of Form

Carol H. Weiss (1997) presented that Evaluation is the methodical assessment of a program or policy's functioning and/or results, in comparison to a set of explicit or implicit criteria, with the aim of enhancing the program or policy.

Michael Quinn Patton (1997) defined that Evaluation is the organized process of ascertaining the value, worth, or significance based on criteria regulated by a predefined set of standards.

Reasons for Evaluation

The demands on Extension programs for enhanced efficiency, effectiveness, and public accountability are on the rise. Evaluation offers a range of approaches to address these pressing needs. Followings are the reasons for evaluation:

1. Planning
2. Analysis of Program Effectiveness or Quality
3. Direct Decision-Making
4. Maintaining Accountability
5. Advocacy

Top of Form

Common Barriers to Evaluation

When it comes to evaluating initiatives or programmes, individuals often exhibit reluctance due to potential disruptions and their impact on outcomes. Several commonly observed barriers to effective evaluation include:

Lack of Management Support: If evaluations are not initiated or backed by high-ranking officials such as the Minister, Deputy Minister, or senior executives, they may not receive the necessary resources and support. Program staff members are often burdened with additional data collection and analysis responsibilities alongside their regular duties, resulting in poorly designed and executed evaluations that provide decision-makers with inaccurate information.

Lack of Skills and Resources: Inadequate or erroneous evaluation results can arise from a dearth of skills, knowledge, and resources in data collection, analysis, and interpretation. Many organizations lack the time, resources, or personnel with the requisite expertise to develop suitable interview or data gathering tools, analyze the collected data, and interpret it accurately. This can lead to flawed assumptions and conclusions based on incomplete, inaccurate, or irrelevant data.

Lack of Relevant Data: The absence of pertinent information renders an evaluation ineffective. If data collection tools and procedures are not carefully designed from the outset of the evaluation process, it may result in missing, inconsistent, or delayed data. The presence of incomplete or inaccurate data regarding certain evaluation questions can render the results meaningless.

Fear of Consequences: The fear that evaluation findings may recommend the elimination or significant alteration of a program can work against the evaluation process. However, it is important to recognize that even a subpar evaluation typically leads to improvements being made to the existing program.

(Anonymous 2021)

Types and Uses of Evaluation

To develop a well-suited evaluation approach, it is essential to comprehend the variations among various types of evaluations. Numerous evaluation designs exist, and the selection of an appropriate one should align with the developmental stage of the program or program activity. The level of effort and the methodologies employed will be contingent upon the program's specific stage and scope.

Evaluation Types	When to use	What it shows	Why it is useful
Formative Evaluation Evaluability Assessment Needs Assessment	During the development of a new program. • When a current program is being altered or is being adjusted in a new population or with a new setting.	Whether the population you want to target is likely to need, understand, and accept the suggested programme components. • How much of an evaluation is feasible based on the goals and objectives.	• It permits for alterations to be made to the plan before completely implementing. • Maximizes the likelihood that the program will succeed.

Evaluation Types	**When to use**	**What it shows**	**Why it is useful**
Process Evaluation Program Monitoring	• Immediate beginningofprogram implementation. • During continuing of an ongoing program.	• How well the program is operating. • The degree to which the programme is being executed exactly as planned. • Whether the program's intended population will find it acceptable and accessible.	• Gives an early warning for any obvious problems. • Enables programmes to keep track of how effectively their activities and strategies are working.
Outcome Evaluation Objectives-Based Evaluation	• When at least one member of the target population has been reached by the programme.	The extent to which the programme is having an impact on the behaviours of the target population.	• Indicateswhether the programme is succeeding in accomplishing its objectives.
Economic Evaluation: Cost Analysis, Cost-Effectiveness Evaluation, Cost-Benefit Analysis, Cost-Utility Analysis	• When a program begins. • During the execution of an ongoing program	• What resources are utilised in a programme and what their direct and indirect expenses are in relation to the results.	• Enables programme managers and funders to compare costs and effects. "How much bang for your buck."
Impact Evaluation	• During the execution of an ongoing program at appropriate intervals. • At the end of a program.	• The extent to which the program achieve its final goal.	• Offers evidence for use in financing and policy decisions.

(Anonymous 2022)

Trends in focus of agricultural extension and implications for evaluation

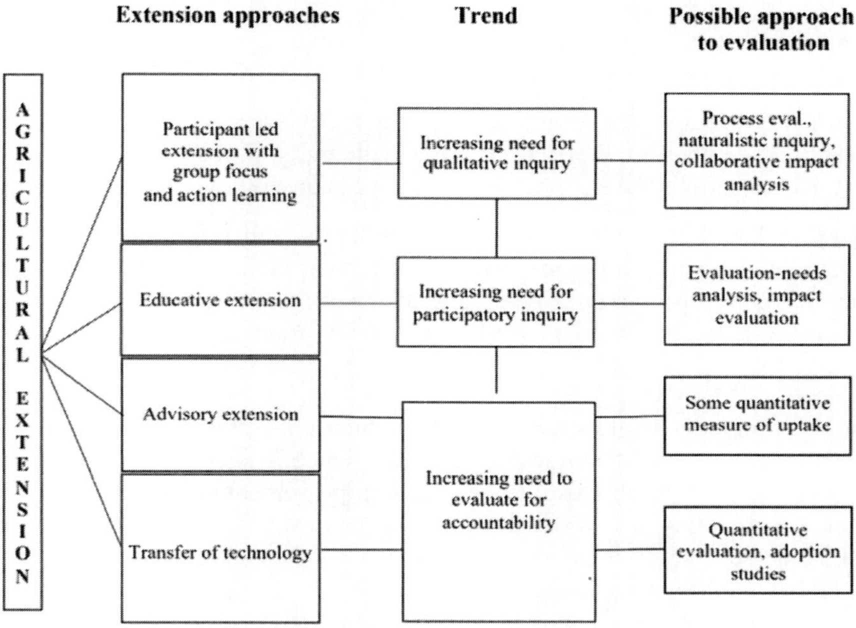

Models of Evaluation

1. Kirkpatrick's Model: Four levels of learning evaluation

The Kirkpatrick Model, a familiar framework for many, facilitates the analysis and evaluation of training and educational endeavors, encompassing both formal and informal instruction. It operates on four distinct levels, making it applicable to all types of training programs (Kirkpatrick, 1959).

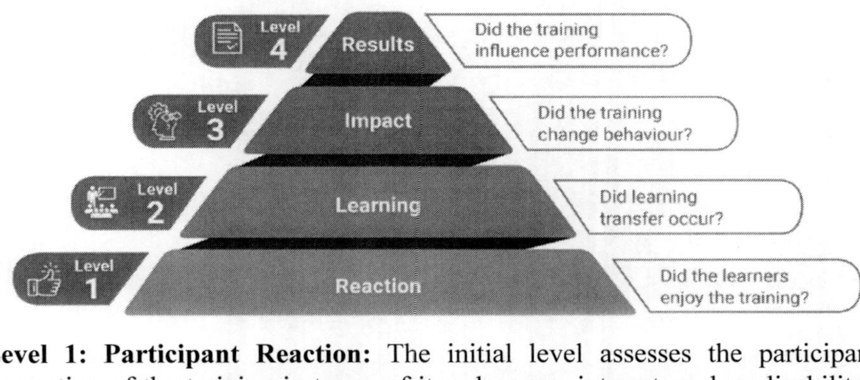

Level 1: Participant Reaction: The initial level assesses the participants' perception of the training in terms of its relevance, interest, and applicability to their work. It focuses on the satisfaction, engagement, and perceived relevance of the training for the learners.

Level 2: Learning: The second level measures the extent to which participants have acquired the intended knowledge, skills, attitudes, confidence, and commitment as a result of the training. It is evaluated through pre- and post-assessments that assess the learners' comprehension and retention of the training content.

Level 3: Behavior: The third level focuses on observing and assessing whether participants apply what they have learned in the training when they return to their jobs. It examines the extent to which behavior changes occur as a result of learning

Level 4: Results: The final level assesses the overall impact of the training, support, and accountability measures on achieving the desired objectives. Level four compares the learning outcomes to the organization's key performance indicators (KPIs) that were established prior to the training.

2. CIPP Model

The CIPP model, developed by Daniel Stufflebeam in the 1960s, is recognized as a decision-oriented framework that systematically collects program data to identify strengths, weaknesses, and areas for improvement in content and delivery (Zhang et al., 2011). This model is commonly employed for management-oriented evaluations due to its integration of four evaluation stages. By focusing on the general aims or mission of a program (Context Evaluation), the plans and resources (Input Evaluation), the activities or components (Process Evaluation), and the results or objectives (Product Evaluation), the CIPP model facilitates continuous development and enhances program effectiveness (Zhang et al., 2011).

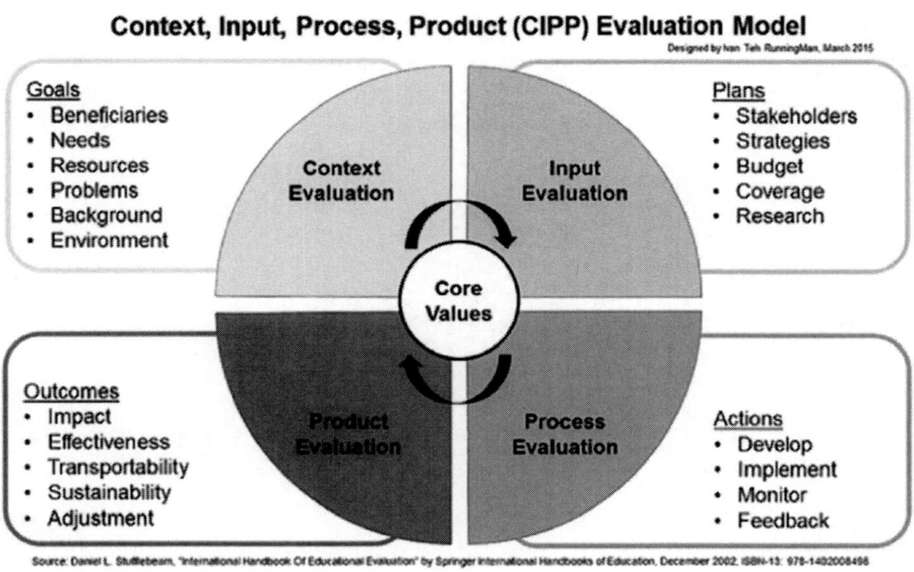

Process for CIPP

The CIPP (Context, Input, Process, and Product) model enables evaluators to navigate through four stages to identify essential elements and determine areas for improvement within programs (Anonymous, 2015). Beginning with context evaluation, evaluators assess program resources, scope, and supports, including overarching objectives, historical data, and cultural context. Once goals have been evaluated, evaluators move to the input evaluation phase, identifying key stakeholders and examining the program planning, implementation strategies, human resources, budget and timeline. The process review stage focuses on evaluating program activities, emphasizing continuous improvement and addressing areas requiring change. Finally, the CIPP model assesses program outcomes and their efficiency, encompassing impact, sustainability, and alignment with the program's mission or fundamental values (Anonymous, 2015; Stufflebeam, 2003).

3. Logic Model

A logic model serves as a visual representation or roadmap that illustrates the interconnected relationships between a program's activities, resources, outputs, and outcomes/impacts (Hayes et al., 2011). It establishes a cause-and-effect or "if-then" relationship between the various components of the program. The logic model helps clarify the distinction between the program's actions ("what" the program is doing) and the desired changes or outcomes ("so what") that are expected to result from the effective implementation of those actions.

The primary purpose of a logic model is to facilitate evaluation by outlining the connections and assumptions between the program's activities and the expected changes (Helitzer et al., 2010). It can be particularly useful in identifying discrepancies between different components of the program, underlying assumptions, and anticipated results.

Logic models can be applied at any level of a business or program, whether it encompasses the entire organization, specific departments or programs, or even particular sections within a department or program. The format and level of detail in logic models can vary significantly. While not all elements are included in every logic model, key components commonly found in logic models include:

Inputs: The resources required for implementing the program's activities.

Activities: The specific actions undertaken by the program and its staff using the available resources.

Outputs: Tangible products, deliverables or capacities that result directly from the program's activities.

Outcomes: The changes that occur in other people or conditions as a direct result of the program's outputs and activities.

Impacts: (Sometimes included) The most distal or long-term outcomes that reflect the ultimate effects of the program.

Moderators: Contextual factors that may influence or affect the achievement of desired outcomes but are beyond the control of the program.

By incorporating these elements, a logic model provides a structured framework for understanding the program's theory of change and aids in planning, implementation, and evaluation processes.

Top of Form

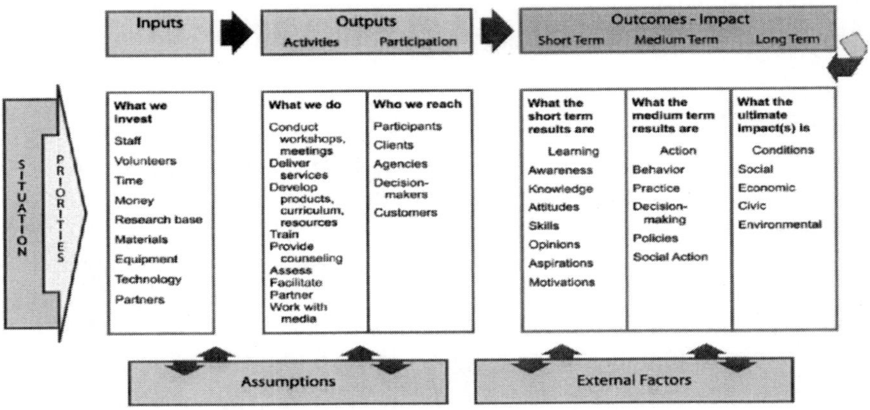

The process of developing a logic model is considered crucial as it compels programs to explicitly state their goals and vision. This fosters a more structured approach to evaluation and establishes a precise hypothesis to be tested.

4. Goal-free evaluation

Goal-free evaluation (GFE) refers to an evaluation approach in which the evaluator conducts the evaluation without prior knowledge or reference to predefined goals and objectives. Introduced by Michael Scriven in 1972, GFE focuses on observing and quantifying actual results, effects, or impacts of a program, regardless of whether they were intended or not. Instead of seeking to understand the program's goals, the GFE evaluator examines what the program actually does and questions what it aims to achieve.

Key Characteristics of Goal-Free Evaluation

- The evaluator actively avoids acquiring knowledge of the program's goals.
- The evaluation is not guided by predetermined objectives.
- The evaluator has limited interaction with program participants or staff.

- By not being aware of the program's goals, the evaluator is more likely to observe unexpected impacts.

Critical Elements of Goal-Free Evaluation

- The evaluator needs to possess expertise in the subject matter being evaluated.
- Evaluator competence is essential, including the requirement for objectivity.

In summary, goal-free evaluation focuses on assessing program impacts without prior knowledge of goals, allowing for an exploration of unintended outcomes and a more objective assessment of program effectiveness.

Difference between Monitoring and Evaluation

Basis for comparison	Monitoring	Evaluation
Meaning	Monitoring involves a systematic and regular process that scrutinizes the activities and advancement of the programme while also identifying any bottlenecks that may arise during its implementation.	Evaluation is an occasional process utilized to reach conclusions about the significance and efficacy of a project or program.
Related to	Observation	Judgement
Occurs at	Operational level	Business level
Process	Short term	Long term
Focuses on	Improving efficiency	Improving effectiveness
Conducted by	Internal Party	Internal or External Party

(Survi, 2017)

Conclusion

The successful implementation of extension programmes relies on the systematic and diligent application of Monitoring and Evaluation practices. Through this chapter, we have explored the interconnection between M&E and its indispensable role in ensuring the effectiveness, efficiency, and sustainability of such initiatives. Monitoring and Evaluation are not standalone processes but are intrinsically linked and must be integrated from the outset of an extension programme. When thoughtfully executed, M&E enhances the overall impact and sustainability of the programme, contributing to the broader objectives of development and progress. As we strive for evidence-based decision-making and transformative change, a strong commitment to Monitoring and Evaluation is essential to unlocking the true potential of extension programmes and realizing a brighter future for all.Top of Form

References

Anonymous. (2015). CIPP evaluation model. Retrieved from http://ivanteh-runningman.blogspot.com/2015/03/cipp-evaluation-model.html

Anonymous. (2021). Information on benefits types evaluation. Retrieved from https://www.policynl.ca/policydevelopment/pages/information-on-benefits-types-evaluation.html

Anonymous. (2022). Program evaluation. Retrieved from http://www.cdc.gov/std/program/ProgEvaluation.pdf

Gosling, L., & Edwards, M. (2003). Toolkits: A practical guide to assessment, monitoring, review and evaluation (2nd ed.). Save the Children, UK.

Hayes, H., et al. (2011). A logic model framework for evaluation and planning in a primary care practice-based research network (PBRN). Journal of the American Board of Family Medicine, 24, 576-582.

Helitzer, D., et al. (2010). Evaluation for community-based programmes: The integration of logic models and factor analysis. Evaluation and Programme Planning, 33, 223-233.

IFRC. (2011). Project/programme Monitoring and Evaluation (M&E) Guide. International Federation of Red Cross and Red Crescent Societies. Geneva, 2011.

Kirkpatrick, D. (1959). Four-level training evaluation model. US Training Development, 13, 34-47.

Patton, M. Q. (1997). Utilization-focused evaluation: The new century text. SAGE Publications.

Scriven, M. (1972). Pros and cons about goal-free evaluation. Journal of Educational Evaluation, 3(4), 1-7.

Scriven, M. (1991). Evaluation Thesaurus. SAGE.

Stufflebeam, D. (2003). The CIPP model of evaluation. In T. Kellaghan, D. Stufflebeam& L. Wingate (Eds.), Springer international handbooks of education: International handbook of educational evaluation.

Stufflebeam, D. L. (1973). Evaluation as enlightenment for decision-making. In B. R. Worthen & J. R. Sanders (Eds.), Educational evaluation: Theory and practice. Worthington, OH: Charles A. Jones Publishing.

Survi, S. (2017). Key Differences between Monitoring and Evaluation. Retrieved from https://keydifferences.com/difference-between-monitoring-and-evaluation.html.

Weiss, C. H. (1997). Evaluation: Methods for studying programs and policies (2nd ed.). Upper Saddle River, N.J: Prentice Hall.

Worthern, B. R., Sanders, J. R., & Fitzpatrick, J. L. (1997). Program evaluation – Alternative approaches and practical guidelines. New York: Longman, Inc.

Zhang, G., Zeller, N., Griffith, R., Metcalf, D., Williams, J., Shea, C., & Misulis, K. (2011). Using the context, input, process, and product evaluation model (CIPP) as a comprehensive framework to guide the planning, implementation, and assessment of service-learning programs. Journal of Higher Education and Outreach Engagement, 15(4), 57-83.

10

Introduction to Market-Led Extension: A Paradigm Shift in Agriculture

Rishabh Singh Gaur*[*1], Soumya Ranjan Behera[2] and Priya Ranjan Mohanty[3]

[1]*PhD Scholar, Department of Agricultural Extension Education, ICAR-NDRI Karnal, Haryana, 132001.*
[2]*PhD Scholar, Department of Agricultural Extension Education, College of Agriculture, Vellayani, Trivendrum, Kerala Agricultural University, Kerala*
[3]*Research Scholar, Department of Agriculture Extension, Palli siksha Bhavana, Visva Bharati University, Shantiniketan*
**Corresponding author's email:* rishabhsg98@gmail.com

Abstract

Indian agriculture has advanced significantly over the past 50 years, becoming self-sustaining in food grains production. Despite a large increase in productivity, the farmers have not been able to be attracted by financial incentives. Farmers must become producers and sellers to get the actual price for their investments in light of the globalization of the commodity market. Farmers are given the majority of production-related technology through an extension delivery mechanism through market led extension. To raise the income of the farmers, it is urgently necessary for the current situation to shift the extension delivery mechanism's emphasis from production to market. Market led extension aids farmers in enhancing the calibre of agricultural output, raising product value and marketability, and improving the farmers' income. This chapter covers concepts, paradigm shifts, advantages and challenges for a better insight to future researchers and policymakers.

Keywords: *Contract Farming Farmer, Farmers Interest Groups, Market Led Extension, SWOT Analysis.*

Introduction

Market led extension effectively delivers sufficient and high-quality information to farmers so they can make decisions about production and marketing that will maximize their return on investment without endangering the needs of future generations. Farmers must improve from being producers or sellers in the home market areas to producers cum sellers in a large market to best reap the benefits of their products. This necessitates the transfer of new extension strategies involving a range from local market to export markets, subsistence to commercial agriculture, productivity to profitability, commodity-oriented to farming systems-oriented, mono-cropping to varieties of crops, and so forth. The extension agents' attention must be diverted from production in some way. Reducing production costs, increasing product value, and enhancing marketability will assist the farming community in receiving high returns for the products.

Goals of market-driven Extension

- To pinpoint potential areas for extension or assistance in agricultural marketing.
- To establish and employ efficient extension approaches for delivering need-based assistance to the agricultural community in the sale of their goods.
- Market connection broadens the current relationship between farmers, researchers, and extension agents.
- To recognize and share innovations in value-adding procedures and post-harvest techniques.
- To create networks and expansion plans to promote the promotion of the product at various levels.

The paradigm shift in Market Led Extension approach

In current agricultural extension scenarios concerning the Indian market, the paradigm shifts from "Production Led Extension" to the "Market Led Extension approach" motivates agriculture and related sectors into profit-oriented businesses the necessary interference based on the market survey and intelligence. To maximize the profits, a farmer must improve from a simple producer on the local market to a producer-cum-seller on the globalized market. Indian agriculture is marked by minimum access to quality information about markets, expensive input costs, shortage of labour, poor storage facilities, market infrastructure and intelligence which led to high marketing costs and a small share of the rupee spent by consumers. Farmers need enough information from when they decide to invest in farming until they start seeing a return. When it comes to producing and selling agricultural products, extension agents who have received training as information-dissemination specialists should be included at all stages of the decision-making process for farmers. This will help them make the best choice at the appropriate moment. Are answers to issues like what to produce necessary for farmers? The place to sell? What are you selling? How will he market his goods?

Paradigm shift factors

Major agricultural development objectives are directly related to the various extension functions. With the changing demands of the rural community, the government attempts to achieve the goals. So, the agricultural extension paradigm shifts from a single to multiple functions. When one objective is met, the government concentrates on accomplishing the following. To suit the evolving requirements of the rural community, extension paradigms likewise evolve during the process. The transfer of technologies from researchers to farmers helps governments and maintains the ultimate goal of a government: national food security.

Market-Led Extension Foundations

a) Product-based implements and tools: It's crucial to understand the many tools, implements, and machines that are offered on the market. For example, there is specialized equipment for planting and harvesting potatoes and sugarcane, among other crops.

b) Current market knowledge: Current market knowledge will assist farmers in making informed decisions about which tools to use and how to best use of such tools.

c) Market intelligence: Numerous service sectors today offer implements based on demand. Consequently, a machine can be rented rather than purchased.

d) Technology use: Using a cell phone, it is possible to remotely start/stop an irrigation pump, which is only one example of how technology is now useful in improving system efficiency. Both extension personnel and farmers must know the existence of such technology.

b) Appropriate extension approaches: Agents should assist the farmers in making decisions for the purchase or rental of any machine or implement. A cost-benefit analysis of employing them should be conducted.

Principles of Market-Led Extension

- Market-based Approach: The foundation of market-led extension is a full grasp of market demand, including the kinds of goods and quality standards needed. Farmers are instructed to create marketable things that conform to consumer desires.

- Farmers-focused It is essential to customize extension services to the unique requirements and goals of particular farmers or farmer groups. This strategy promotes active involvement and process ownership.

- Collaboration and Partnerships: Collaboration amongst a variety of players, including government agencies, NGOs, businesses, and farmer organizations, is essential for market-led extension to be successful. These collaborations may offer a variety of resources and skills.

- Innovation and technology: A key component of market-led extension is incorporating novel techniques and technologies. Farmers are urged to adapt contemporary farming methods, use data to influence decisions, and access information through digital channels.
- Business and Financial Literacy: Farmers must have a business and financial management abilities in order to be market-savvy. Training on marketing, financing access, and budgeting should be part of extension programs.

Prospects of Market-led Extension

Because of the current global challenges in food production, Market Led Extension holds significant potential for facilitating sustainable and efficient food production practices. 'Lab to Land' has been heavily emphasized in our nation over the years; now, the emphasis should be on 'farm to fork. Because of the World Trade Organization (WTO), the world market is open to all buyers and sellers. Producers are eager to export their excess goods anyplace in the world with a competitive advantage. Therefore, in the future, Market Led Extension will take the main responsibility for agricultural development in the Indian scenario (Clarke, 2000). Creating farmers' groups based on commodity, linkage between them, improving farm management ability, farmer exposure and advice for making a quality market decision are all included in Market Led Extension. Since it is the primary responsibility, the key is to improve farmers' socio-economic condition.

Strategies for Market-Led Intervention Expansion

In Market-Led Extension, the combination of various sectors such as extension education, agricultural economics, and agricultural marketing have a crucial role. Market led extension aims to enhance the farmers' condition through low production costs and great quality product productivity with the help of Farmers Interest Groups (FIGs). Creative initiatives should be ingrained in the extension system to access the market knowledge of the clients and a secure market for their agricultural products. One of these tactics is the timely introduction of commodity interest groups (CIGs), and FIGs for marketing agricultural products. In the end, this will allow for the participation of different stakeholders in the sale of agricultural products, maximizing their opportunity for maximum profit. Utilizing the idea of e-extension for marketing objectives is yet another successful extension method. An e-commerce system can achieve this through the involvement of farmers, traders, wholesalers, and financial transactions.

Additionally, reducing the difficulties associated with post-harvest loss can be accomplished largely by implementing recovery funds like subsidy. In addition to establishing marketing agro-processing links and market intelligence, extension personnel will need to be prepared to carry out a number of additional tasks, such as conducting a SWOT analysis of the market, supporting and enhancing the capacities of locally established groups, organizing Farmers Commodity Interest Groups, etc., creating websites of progressive farmers (Paresh et al., 2012).

Implementation of Market-Led Extension

- Needs Assessment: The procedure starts with carefully examining the farmer population's requirements, opportunities, and limitations. Knowledge, skill, and resource shortages are revealed through this examination.
- Tailored Training: Extension professionals give farmers specialized instruction and advice based on their needs. This could include marketing tactics, post-harvest handling methods, or agronomic advice.
- Information Access: Having access to current market information is essential. The internet and mobile technology are useful resources for spreading market pricing, weather predictions, and best practices.
- Value Chain Analysis: Farmers can find opportunities for value addition and create competitive advantages by understanding the full value chain, from production to consumption.
- Monitoring and assessment: Ongoing monitoring and assessment are important for determining how well extension programs work. The extension services are modified and improved in response to farmer feedback.

Advantages of Market led extension

- Higher Incomes: by matching farming practices to consumer wants, farmers are more likely to obtain higher prices for their produce and raise their overall income.
- Market-led Extension encourages sustainable farming methods that preserve natural resources and safeguard the environment.
- Rural Economic Growth: the strategy helps rural economies grow by bolstering agricultural value chains and generating employment opportunities.
- Resilience: Farmers are better equipped to adjust to shifting market conditions and problems brought on by the climate when they have knowledge and market insights.

Enhanced duties for extension personnel in light of market-led extension

- SWOT analysis of the market: Markets need to be examined for their various factors like strengths (market demand, high price, etc.), weaknesses (the opposite of the aforementioned), opportunities (export facility, selling time etc.), and threats (quality and perishability of the produce, etc.). Therefore, The farmers must be informed of this analysis to plan production and marketing.
- Assisting and strengthening the capabilities of locally founded organizations participating in various programs and schemes, such as watershed committees, user's groups, SHGs, water users' associations, and thrift and credit organizations. These groups must be aware of self-help activity's significance, usefulness, and advantages.

- Educating the agricultural community so that they will view agriculture as an entrepreneurial activity and organize the various stages of crop production and commercialization accordingly.
- Farmers should be made aware of the advantages of direct marketing. Rytu Bazars in Andhra Pradesh, Apni Mandis in Punjab and Haryana, and Uzavar Santhaigal in Tamil Nadu have demonstrated success in some states.
- Building the capacity of FIGs to improve production, post-harvest operations, storage, transport, and marketing
- Regularly acquiring thorough market intelligence on a variety of market-related topics.
- Regular computer use of the internet to stay informed about market intelligence.
- Publication of agricultural market information in newspapers, radio, television, and the internet.
- Making videos featuring farmers who have succeeded in growing a particular crop.
- Development of websites for successful FIGs in the field of agricultural management that contain all the necessary data to aid other FIGs in succeeding

Challenges Facing Agricultural Extension Staff in Light of Market-Led Extension

For extension employees, knowledge of adequate farm machinery is essential. Thanks to this, the farmer will be assisted in making cost-effective tool/implement choices. They ought to try to connect farmers and retailers of farm equipment. They should frequently ask about the newest machines that are on the market. They shouldn't allow the store to influence them and enforce farming equipment that is not cost-effective for the farmer. In order to determine the suitability of a machine, either to be purchased for a single farmer or a group of farmers, extension professionals should study the crop coverage area. A combine harvester, for example, costs between 10 and 15 lakh rupees. The amount of wheat that can be harvested in the area should be investigated, as well as whether farmers are willing to pay for the equipment's rental. Before investing on a pricey machine, they should consider economic math. There are further difficulties, which are detailed below:

a) The difficulty is in inspiring the extension staff to learn new marketing techniques and skills and information about pertinent farm machinery.

b) A vast network of knowledgeable individuals must be involved in gathering recent data and need dedicated websites to publish according to farmers needs.

c) They must know several custom hiring models appropriate for farmers.

d) It's important to update the necessities, release the extension cadres from bureaucratic restrictions, and encourage them to achieve motivational elements like success, happiness, job, and acknowledgment.

Methods for reorienting agricultural extension staff towards market-led extension

- Identification of the farmer or group of farmers interested in exporting agricultural products.
- Spreading knowledge about market-driven extension among farmers.
- Teach farmers how to reduce their costs by using cost-saving techniques.
- The establishment of model farming practices for export.
- Providing farmers with knowledge about finances and the market.
- Scheduling grading and packing training sessions for the farmers.
- Disseminating information via online sources, television, and other electronic media.

Functions of Contract Farming in Market-Led Extension

New farming systems and models are emerging as a result of India's agriculture's rapid transition, promising farmers a stable market and a fair price for their produce. Contract farming will likely play a significant part in this shift, which is becoming clearer. The National Agriculture Policy of the Government of India also states, "Private sector participation will be promoted through contract farming and land leasing arrangements to allow accelerated technology transfer, capital inflow, and assured the market for crop production, especially of oilseeds, cotton, and horticultural crops." The idea of contract farming has evolved to connect productive agriculture to the market. It promises to establish an appropriate connection between "farms and market."It can be a way to draw attention to Indian agriculture on the market. The key component of this arrangement is the producer/seller's promise to deliver an agricultural product of a specific quality, in the quantity needed by a known and committed buyer, at the price and time agreed upon. The fundamental components of contract farming are pre-determined pricing, agreed-upon quality, quantity, and timely delivery of produce. A system for producing and supplying land-based and related products by farmers/primary producers under advance contracts is known as contract farming (CF). The essence of such agreements is a commitment to provide an agricultural commodity of a type, at a specified time, price, and in specified quantity to a known buyer (FAO claims that Contract).

Benefits of Contract Farming

For farmers

The sponsor frequently provides the necessary inputs and manufacturing services through credit system.

- New technology is frequently introduced through contract farming, and farmers can gain new skills in the process.
- Farmers' pricing risk is frequently lessened because many contracts have fixed prices.
- Contract farming can provide small farmers access to new markets they wouldn't otherwise have.

Sponsors only

- Production on estates, for instance, is less politically palatable than contract farming with small farmers.
- Land limits can be overcome by working with small farmers.
- Because production is not the sponsoring company's responsibility, less risk is involved compared to open-market purchases.

Farm mechanization in relation to market-led extension

Machines, tools, and implements are important agricultural inputs. Mechanizations typically employed an overall description of how these inputs were applied. These tools, devices, and equipment are powered by three types of farm power: manual (human), animal draft, and motorized. People provide up to 80% of the farm power in many developing nations. However, most of the time, machines are used as the power source in developed nations. Advanced crop husbandry inputs, such as better seeds, fertilizer, pesticides, and higher use of irrigation water, cannot be implemented in the Indian environment, where the growth of agricultural areas is constrained. Small and marginal land holdings (less than 2.0 ha) make up 85% of all land holdings in India, where the average farm size is a meagre 1.16 ha (Mehta et al., 2014). Increased farm power can directly enhance production in circumstances where land is not a restricted by merely expanding the amount of land or animals that one man can manage. Agriculture is immediately impacted by changes in population expansion, urbanization, and market dynamics, making it more competitive and market-oriented.Farmers must adapt their skills and competencies to deal with these shifting farming settings due to these changes. Extension personnel are necessary to help farmers manage and operate their farms more profitably. However, because their knowledge and practice have mostly focused on agricultural output and technology transfer, many extension workers find that managing a business is problematic.

Issues and Challenges of Market-Led Extension

The scale of the public extension system is enormous, and it is overworked with several tasks. Now that a new marketing component will be added to the system, there may be a number of challenges to overcome. Again, before engaging in marketing extension, extension personnel must be driven to acquire various marketing abilities. The greatest difficulty for market-led extension will be to develop accurate, pertinent, thorough, and timely market intelligence about products, market practices, customers, and commodity prices (Roy, 2019). An information policy needs to be defined to provide farmers with improved websites and a wealth of information. Additionally, there are some glaring difficulties with the disparity of storage, warehouse, and transportation facilities at various locations.

Although there is a sizable and growing market in agriculture and related industries, the agricultural marketing system cannot keep up with the market's growth. 85 percent of the 27,294 rural periodic markets lack the necessary infrastructure for efficient commerce, and 7161 of those markets are run improperly (Shitu et al., 2013). Due to inadequate storage facilities and a weak marketing chain, corers are lost on average at Rs. 50,000 per year. The development of an effective and efficient marketing system is hampered by the farmers' inability to assume risk in the production and marketing of agricultural products, the absence of storage and cold-chain infrastructure for farm products, the absence of grading and standardization, and the absence of agro-processing facilities. The extension system is enormous in scope and extensively taxed by numerous endeavors. The growing communication gap between the client and research systems only worsens it. Another difficulty is gathering reliable market intelligence or information. Comprehensive, accurate, timely, pertinent, private, reliable, and equally accessible are all qualities of good market intelligence. Due to their perishable nature, agricultural products are considerably distinct from other marketing items. Because manufacturing is seasonal, supply is irregular. Small land holdings among farmers result in sporadic production and a range in product quality. Furthermore, due to untapped potential in agricultural marketing and underutilized export chances for agricultural products in our nation, our country is unable to gain from the WTO through export.

Conclusion

Farmers are now able to receive high value for their crops, reduce investments in production, and increase the value and marketability of their products because of market-led extension. To provide farmers with information about production and markets, it is necessary to fully utilize the possibilities of electronic, print, and information communication technology. The emphasis must be changed from being supply driven to market driven. To distribute knowledge about production, but mostly marketing-related information, extension agents need to work dedicatedly in the rural communities for marketing by employing extension techniques. This can be accomplished by giving farmers the tools they need to get the best possible returns on their produce. As a result, the production costs will be reduced, and the products will be improved through value addition. As a result, the market-led extension system can be used to promote sustainable agriculture. The extension functionaries' attention needs to go beyond production. This will lower production costs, increase product value, and enhance the marketability of the products for the farming community.

References

Ahmad, N., and Slathia. (2011). Market Led Extension for Promoting Rural Entrepreneurship in India. Journal of Global Communication, 4(2): 143-147.

Clarke L. J. (2000). Strategies for Agricultural Mechanization Development: The Roles of the Private Sector and the Government. Agricultural Support Systems Division, FAO Publication, Rome, Italy.

Mehta C.R., Chandel N.S. and Senthilkumar T. (2014). Status, challenges and strategies for farm mechanization in India. Agricultural Mechanization in Asia, Africa, and Latin America, 45(4):43-50.

Paresh, V.J.., Milind, B. and Sarita, B. (2012). Marketing Decision Model: A Market Led Extension technique for Efficient Agricultural Marketing. Indian Streams Research Journal, 2(3): 1-4

Roy, H. (2019). Market Led Extension in Agricultural Development. Agriallis, 1(1): 26-28. https://agriallis.com/wp-content/uploads/2019/11/Market-Led-Extension-In-Agricultural-Development-.pdf

Shitu, G. A., Sakia, A. R., Meti, S. K. and Maraddi, G. N. (2013). Market led extension; prospects and challenges for Agricultural sustainability in the 21st century. Conference Paper.

11

Empowering Indian Agriculture: Farmer Producer Organizations (FPOs) as Catalysts for Rural Transformation

Rebba Kishore Kumar*, Pooja Mohanty, Kiran Sourav Das and Meesala Haritha

Research Scholar, Department of Agricultural Extension Education
Odisha University of Agriculture and Technology, Bhubaneswar, Odisha
Research Scholar, Department of Extension and Rural Sociology
Tamilnadu Agricultural University
**Corresponding author's e-mail: kishorerebba81@gmail.com*

Abstract

The agricultural sector in India, often referred to as the backbone of the nation's economy, faces multifaceted challenges. Small and marginal farmers, constituting a substantial majority, grapple with limited resources and formidable transaction costs that hinder their participation in modern agricultural markets. To address these constraints and leverage the potential of smallholder farmers, Farmer Producer Organizations (FPOs) have emerged as a promising institutional intervention.This comprehensive overview explores the evolution, principles, formation, and operational framework of FPOs in India. With a surge in FPO registrations over the past two decades, particularly after 2014, these organizations have gained momentum across states. Crucially, FPOs face financial challenges, with a skewed distribution of paid-up capital and a substantial percentage of non-compliant organizations. Despite these hurdles, FPOs have been instrumental in providing a wide array of services to their members, ranging from input supply and marketing to financial and training services. This abstract provides a glimpse into the rich tapestry of FPOs in India, highlighting their vital role in enhancing the economic well-being of small and marginal farmers and contributing to rural development. It underscores the need for continued support, both from governmental and non-governmental entities, to ensure the sustainable growth and success of FPOs in the Indian agricultural landscape.

Keywords: *Agriculture, Development, Producer Organization, Resources, Small Farmers.*

Introduction

Agricultural sector is the pillar of the Indian economy. It provides employment to around half of the workforce of the country and contributes around 18 per cent to the gross value added of the country. Small and marginal landholdings, constituting less than 2 hectares, make up a significant 85 percent of India's farming community (Joshi, 2022). These small farmers encounter various challenges, including farm-specific and household transaction costs that limit their participation in input and product markets. With changing food preferences driven by income and population growth, there's a growing need for small-farm commercialization to meet the rising demand for diversified, higher-quality diets (Chakraborty, 2020). To address the constraints faced by small farms, aggregation models have emerged as institutional interventions.

In contrast, Farmer Producer Organizations (FPOs) have emerged as an alternative to cooperatives in India, following the Companies amendment Act in 2013 (Kumar et al., 2022). FPOs represent a unique blend of private enterprise and cooperative principles. They have adopted a hybrid business model that distinguishes them from traditional cooperatives. FPOs maintain exclusive membership for primary producers who contribute fairly to the organization's working capital. Importantly, they operate under a democratic framework, ensuring that every member has equal voting rights. This distinct structural approach sets FPOs apart from cooperatives, with a primary mission to facilitate the commercialization of small-scale farming and uplift farm incomes. The central government's strong push has led to the registration of thousands of FPOs over the past decade. However, there is a need to gain a deeper understanding of their performance and viability, including their geographical distribution, membership patterns, and financial attributes.

History of FPOs in India

Producer Organizations are primarily described as "membership-based entities or federations of organizations with elected leaders accountable to their constituents." They serve as legal entities formed around primary products, with a major membership base comprising farmers, fishermen, milk producers, and artisans. These organizations can take various forms, such as cooperative societies, producer companies, or other entities aimed at promoting resource-sharing and benefits among their members. The Department of Agriculture & Cooperation, Ministry of Agriculture & Farmers' Welfare, Government of India, has specifically recognized farmer producer organizations registered under the special provisions of the Companies Act, 1956.

Farmer producer organizations represent the collective efforts of producers, particularly small and marginal-level farmers, to address the numerous challenges in agriculture, such as improving access to investments, technology, and input markets. It provides a wide range of facilities to its regular members such as input supply, financial, technical, insurance, procurement, packaging, marketing and networking (Tiwari, 2021). The first registered farmer producer organization in India under these new provisions was the Vanilla India Producer Co. in Kerala, established in 2004. The primary goal of Farmer Producer Organizations was to mobilize farmers into member-owned producer organizations to enhance the productivity and profitability of small farmers.

Categorization of farmer producer organizations

Broadly, farmer producer organizations can be categorized into four types such as Input-oriented, market-oriented, extension-oriented, and policy advocacy-oriented (Gebremedhin, 2012). The specific type of farmer producer organization can vary based on the context, origin, policy trends, and economic changes, and may even shift between these categories during their existence. SFAC, designated by the Department of Agriculture & Cooperation, Ministry of Agriculture & Farmers' Welfare, serves as a single-window organization for fulfilling various training needs and technical support for farmers. SFAC actively promotes FPOs through accredited Resource Institutions (RIs), which provide various inputs related to training, capacity-building, technology, and market linkages, connecting these bodies to input suppliers. The National Cooperative Development Corporation (NCDC) also includes farmer producer organizations in the list of institutions supporting different agricultural development programs in the country. Similarly, the National Agricultural Cooperative Marketing Federation (NAFED) has recognized farmer producer organizations as eligible institutions for working under the price support policy. NABARD has initiated several programs to support producer organizations, adopting a flexible approach to meet the needs of producers, and providing financial support to these organizations since 2011 through the Producer Organisation Development Fund (PODF). The distribution of FPOs across states in India varies widely. Six states, including Maharashtra, Bihar, Odisha, Tamil Nadu, West Bengal, and Uttar Pradesh, collectively account for two-thirds of the total FPOs in the country. Maharashtra stands out, contributing one-third of all registered FPOs, owing in part to its successful history of agricultural cooperatives and structural marketing reforms. Several states, such as Bihar, Odisha, Tamil Nadu, West Bengal, and Uttar Pradesh, have reported a significant increase in FPO registrations in recent years. Govil et al., (2020) reported that till March 31, 2019 total 7,374 Producer Companies (PCs) registered in India.

Structure of a Farmer Producer Organization (FPO)

The structure of a Farmer Producer Organization (FPO) typically follows a hierarchicalmodel that includes various levels of governance and management (Manaswi, 2019):

1. General Body (GB): The General Body is the highest decision-making body of the FPO. It consists of all the member farmers who are part of the organization.

The General Body holds periodic meetings to discuss and decide on key issues, including the election of representatives to the Executive Body and the Board of Directors. It plays a crucial role in setting the overall direction and policies of the FPO.

2. Executive Body: The Executive Body is responsible for day-to-day operations and decision-making within the FPO. It typically includes two representatives from each Farmer Interest Group (FIG). FIGs are sub-groups within the FPO that represent specific farming interests or sectors, such as crop-based groups or livestock-based groups. The Executive Body ensures that the FPO's activities align with its objectives and the needs of its members.

3. Board of Directors: The Board of Directors is a governing body that oversees the FPO's strategic direction, financial matters, and major decisions. It is composed of elected representatives from the Executive Body. The Board plays a pivotal role in ensuring transparency, accountability, and effective management of the FPO.

4. General Manager: The General Manager is responsible for the overall management and administration of the FPO. This role involves coordinating activities, implementing decisions taken by the General Body and Board of Directors, and managing the FPO's resources. The General Manager is often a key point of contact between the FPO and external stakeholders.

5. FPO Staff: The FPO may employ staff members to handle various functions, such as marketing, finance, administration, and extension services. These staff members support the day-to-day operations and service delivery of the FPO. Their roles and responsibilities vary depending on the specific needs and activities of the organization.

6. Local Resource Persons: Local Resource Persons are individuals with expertise in agriculture, technology, or other relevant fields. They may work closely with the FPO to provide technical assistance, training, and advisory services to member farmers. These individuals contribute to the FPO's capacity-building efforts and help improve agricultural practices among members.

Challenges Faced by FPOs

FPOs encounter several challenges in their operations and organizational development. These challenges include organizing producers effectively, delivering agricultural services, capacity-building for staff and board members, capital acquisition (including working capital), staff recruitment and retention, establishing efficient processes and systems for collection, processing, value addition, marketing, and understanding complex value chains for effective interventions.

Operational Framework

The operational framework of Farmer Producer Organizations (FPOs) involves several crucial steps in their promotion and development:

1.Cluster Identification: The process begins with the selection of cluster areas, typically carried out in collaboration with resource institutions and state government departments. These clusters should ideally consist of 8,000 to 10,000 farmers within one or two blocks, covering 80 to 120 adjacent villages in a specific district.

2.Diagnostic Study: Resource institutions conduct a diagnostic study in the selected cluster area. This analytical study assesses the initial situation of farmers and the state of agriculture in the area. It helps identify the specific context for project implementation and the necessary interventions required.

3.Feasibility Analysis: Resource institutions conduct a feasibility analysis to evaluate the viability of forming FPOs in the cluster. This analysis covers various aspects, including technical, political, financial, legal, environmental, socio-cultural, and resource feasibility. The goal is to make a compelling case for establishing FPOs in the specific regional context.

4.Baseline Assessment: Baseline assessment involves gathering data on the current situation of small, marginal, and tenant farmers in the cluster. This assessment identifies potential interventions for development and business growth. It also establishes baseline data for future outcome measurement, typically conducted through household-level interviews and focus group discussions.

5.Business Planning: Business planning is a collaborative process involving selected farmer representatives. A detailed business plan is developed, encompassing various aspects and projections. The plan is created with the active participation of at least 10 percent of FPO farmer members to ensure a clear vision.

6. Mobilization of Farmers: The mobilization of farmers is a critical step in FPO development. A spearhead team, along with a group of farmers, works to mobilize farmers into Farmer Interest Groups (FIGs) and eventually as farmer-members of FPOs. This mobilization is achieved through various communication methods, including village meetings, documentaries, pamphlets, and posters.

7. Organizing & Formalizing: FIGs in the aggregated cluster come together to form FPOs. Typically, around 50 to 70 FIGs can collectively establish an FPO. The formal registration of FPOs is done under the provisions of the Producer Company Act. It's important to emphasize that the process should not be rushed, and the decision to register an FPO should be made by the farmer interest groups at the right time.

8. Resource Mobilization: Before commencing FPO activities, resource institutions, with the assistance of FPO representatives and the board of directors, mobilize various resources. This includes human resources (staff), financial resources, physical assets, and technical resources. Financial linkages with various agencies are established based on the business plans to secure the necessary resources.

9. Systems Development: Developing a robust management system is crucial for FPO promotion. Guiding principles for management systems cover various aspects, including input and output management, financial services, human resources, inventory management, quality control, internal audit, marketing, and conflict resolution. Standard operating measures are established to ensure smooth operations.

10. Business Activities: FPOs begin their business activities, which include procurement, production, processing, marketing, and financial services, among others. Resource institutions play a role in guiding both the operational and governing structures of the FPO to ensure seamless business operations.

11. Performance Appraisal: Constant appraisal of the performance of stakeholders, including farmer members, service providers, and the governing board of directors, is essential. This evaluation helps FPOs identify areas for improvement. Accounting audits and internal processes are established to maintain accountability and transparency, serving as key institutional frameworks for FPO evolution.

These steps collectively contribute to the development and success of FPOs, ensuring they effectively serve the interests of small and marginal farmers while promoting sustainable agricultural practices.

Services Offered by FPOs

FPOs offer a wide range of services to their members, covering various aspects of agriculture and farm management. These services are designed to enhance the productivity and profitability of small and marginal farmers. FPOs provide end-to-end solutions, including input supply, technical guidance, processing, and marketing. Here is an overview of the services offered by FPOs:

1. Organizational Services: FPOs start by organizing farmers into informal groups (Farmer Interest Groups or FIGs). This collective action helps build their capabilities and establish internal monitoring systems, fostering collaboration among members.

2. Production Services: FPOs facilitate collective production activities by forming clusters of farmers to produce a significant volume of marketable surplus. They also suggest appropriate farming systems tailored to specific circumstances.

3. Input Supply Services: FPOs provide member farmers with access to high-quality inputs at competitive prices. This includes seeds, fertilizers, sprayers, pump sets, pesticides, accessories, and pipelines, ensuring that farmers have the necessary resources for cultivation.

4. Marketing Services: After the procurement of agricultural produce, FPOs offer direct marketing channels. This helps members save time, reduce weighment losses, minimize transaction costs, avoid distress sales, manage transportation, mitigate price fluctuations, and maintain product quality, ultimately improving marketing outcomes.

5. Financial Services: FPOs, with support from organizations like NABARD, make loans available to members for various purposes, including crop cultivation, purchase of agricultural machinery like tractors, construction of wells, pump sets, and pipeline installation.

6. Procurement and Packaging Services: FPOs engage in the procurement, storage, value addition, and packaging of agricultural produce sourced from their member farmers. This adds value to the products and ensures better returns.

7. Technology Services: FPOs provide guidance to farmers on best practices, market information, post-harvest processing, and the responsible use of quality inputs. They support farmers in adopting modern and sustainable agricultural techniques.

8. Insurance Services: FPOs offer various insurance services, including coverage for electric motors, crop insurance, and life insurance. They assist farmers in accessing government grants for rural welfare programs.

9. Training and Extension Services: FPOs identify the training and extension needs of their members and provide training on good agricultural practices, business skills, low-cost and environmentally friendly inputs, production techniques, and more.

10. Value-Added Services: FPOs employ various techniques to add value to farm produce locally, thereby increasing profitability for their members.

11. Welfare Services: FPOs contribute to improving the health and safety of their member farmers and assist in resource management, such as water, fisheries, pasture, forests, and soil conservation.

12. Market Linkages: FPOs play a pivotal role in coordinating the supply and demand of agricultural produce by establishing linkages between farmers, processors, traders, and retailers. They also provide key business development services, including access to input suppliers, transport services, and market information, to enhance both profitability and the welfare of their members.

13. Information and Linkages: FPOs offer market price information, product specifications, and access to various business services for rural producers. They help build linkages with financial institutions, processors, traders, consumers, and government programs to further the interests of their members.

In essence, FPOs serve as holistic support systems for small and marginal farmers, addressing their diverse needs and promoting sustainable agricultural practices. These services empower farmers, improve their livelihoods, and contribute to the overall development of rural communities.

Some Major Farmer Producer Organizations

1. Indian Organic Farmers Producer Company Ltd

Location: Aluva, Kerala

Description: Situated in Aluva, Kerala, the Indian Organic Farmer Producer Company Ltd is a cooperative that specializes in the production of organic goods. To become a member, growers are required to hold organic certification. The cooperative charges a patronage fee of Rs. 40,000 per share. Each share enables stakeholders to offer organic products valued at up to Rs. 40,000. Regardless of the number of shares an individual owns, they are entitled to only one voting right.

2. Vanilla India Producer Company Ltd (VANILCO)

Description: VANILCO is an initiative by the Indian Producers Movement (Infarm), a Kerala-based nonprofit organization with a substantial membership of farmers. VANILCO's primary mission is to safeguard the long-term interests of vanilla farmers. It collaborates closely with vanilla farmers to cultivate and market premium-quality vanilla beans and extracts.

3. Chetna Organic Agriculture

Description: The Chetna Organic & Fair-Trade Cotton Intervention Program was launched in 2004 as a comprehensive pilot program focusing on organic and non-pesticide management (NPM) practices. This initiative was a response to India's agrarian crisis. The program aims to enhance the sustainability and profitability of farmers' agricultural systems while also providing access to ethical and fair-trade cotton markets. Its overarching goal is to expand the range of livelihood opportunities for smallholder farmers.

4. AMUL

Description: The Gujarat Cooperative Milk Marketing Federation, popularly known as AMUL, contemplated a transition in its legal structure to become a producer business in May 2006. While the conversion hasn't taken place yet, the idea appears to be under consideration. The initiative to shift AMUL from a cooperative society to a producer corporation was initiated in May 2006, aiming to free individuals from the constraints associated with cooperatives.

5. The Junagarh Dairy

Description: Guided by the Development Support Centre in Ahmedabad, a producer company has been formed in Dari, situated in the Amreli District. This initiative brings together ten watershed development groups, each hailing from a distinct village. Their collective endeavor focuses on providing agricultural inputs and advanced technology to support farming activities.

6. Samarth Kisan

Description: Samarth Kisan, with the highest capitalization among DPIP producer companies, provides inputs and engages in wheat and soybean seed production. In 2008, they initiated an enhanced soybean system, backed by the Solidaridad's Soy Producer Support Initiative (SOYPSI). The primary objective was to enhance the livelihoods of farmers and address environmental concerns.

7. Sagar Samridhi Crop

Description: This corporation's main activities include input supply and the production of soybean and wheat seeds. They have agreements with 20 member farmers to produce seeds each season and own seed processing machinery.

8. Khajurao Crop

Description: Khajurao Crop, a cooperative spanning 99 villages with 140 CIGs, primarily focuses on delivering agricultural inputs and seed production. It purchases seeds from 100 farmers each season.

9. Nowgoan Crop

Description: Nowgoan Crop is involved in seed production and input supply. It purchased land to cultivate vegetable seedlings and selects 30 members each season to produce seeds.

10. Ramraja Crop

Description: Ramraja Crop engages in the production of wheat, soybean, gram, and urd seeds, along with input provision. Farmers with at least 1 hectare of land and a reliable reputation are chosen to produce seeds, and they receive input subsidies.

These organizations, each with its unique objectives and initiatives, play pivotal roles in promoting sustainable and ethical agricultural practices across India (Kumari, 2022).

Conclusion

In conclusion, Farmer Producer Organizations (FPOs) have emerged as a critical driver of change in India's agricultural landscape. With a focus on empowering small and marginal farmers, these entities have rapidly gained traction, especially in the last few years, thanks to favorable government policies and support. FPOs are not just organizations; they represent a transformative approach to agriculture, guided by principles of inclusivity, autonomy, and economic participation. FPOs are not only improving the livelihoods of small farmers but also contributing to the overall well-being of rural communities. However, challenges persist, including compliance issues, varying levels of capitalization, and the need for sustained support during the early stages of FPO development. It is essential to address these challenges comprehensively to ensure the continued success and sustainability of FPOs.

References

Chakraborty, M (2020). Ecosystem for Sustainable Development of Farmers' Producer Organisation. Agriculture & Food:e-Newsletter, 2(9): 478-483. ISSN: 2581-8317.

Gebremedhin, B., Jemaneh, S., Hoekstra, D. & Anandajayasekeram, P. (2012). A guide to market-oriented extension services with special reference to Ethiopia. IPMS (Improving Productivity and Market Success) of Ethiopian Farmers Project. Nairobi: ILRI.

Govil, R., Neti, A., & Rao, M. R. (2020). Farmer Producer Companies: Past, Present and Future. Azim Premji University, Bangalore.

Joshi, E. (2022). Assessing India's FPO Ecosystem. Cornell University News. Retrieved from https://tci.cornell.edu/?blog=assessing-indias-fpo-ecosystem

Kumar, R., Kumar, S., Pundir, R.S., Surjit, V. and Rao, S. (2022). FPOs in India: Creating Enabling Ecosystem for their Sustainability. ICAR-National Academy of Agricultural Research Management, Hyderabad, India. pp22.

Kumari, B. (2022). List of Major FPOs in India. Krishijagran. Published on 15 December, 2022. Retrieved from https://krishijagran.com/blog/list-of-major-fpos-in-india/

Manaswi, B. H., Kumar, P., Prakash, P., Anbukkani, P., Kar, A., Jha, G. K., Rao, D.U.M. & Lenin, V. (2019). Evaluation of farmer producer organizations of telangana: A swot analysis approach. Journal of Community Mobilization and Sustainable Development Vol. 14(3), 457-466.

Tiwari, N., & Upadhyay, R. (2021). Farmer Producer Organizations: Concept, Structure and Functioning. TECHNOFAME- A Journal of Multidisciplinary Advance Research, 10(2): 104-113. https://technofame.com/CMS/Journal/Journal_19/290.pdf

12

Overview of Participatory Extension Approaches in Indian

***Srabani Ghosal**[*1]**, Indira Priyadarsini Pattnaik**[2]* ***and Dorisa Narzary**[3]*

[1,2]PhD Scholar, Department of Agricultural Extension Education, Orissa University of Agriculture and technology (OUAT) Bhubaneswar, Odisha
[3]Research Scholar, Department of Agricultural Extension Education, Dr. Panjabrao Deshmukh Krishi Vidyapeeth , Akola, College of Agriculture Nagpur
[*]Corresponding author's e-mail: srabani.5ghosal@gmail.com

Abstract

The participatory approach has gained importance in agriculture and rural development as it offers inclusive and dynamic methods of engaging local communities in sharing knowledge and decision-making process. Various participatory extension approaches keep farmers and local people in heart for developmental activities that build trust and solve various problems. Participatory approach is a collaborative process that helps exchange innovation, knowledge and adaptive learning that helps in successful interventions. The participatory approach is very useful in bridging gaps between researchers, extension agents and local people to exchange knowledge. Participatory extension approaches highlight the importance of gender equity and social inclusion and ensure that marginalised groups have their voice in developmental processes. There may be various challenges, such as resource requirements and commitments given to local people. However, the benefits of participatory approaches outweigh those that help achieve goals for sustainable agriculture and rural development. Hence, participatory approaches play an important role in increasing resilience, empowering communities and sustainable development.

Keywords: *Approach, Efficiency, Empowerment, Participatory, Self-reliance.*

Introduction

There is an emergence of a transformative paradigm of participatory approaches in various fields such as management, health care and education, community development, etc. Participatory approaches challenge the top-down approach used frequently for decision making (Narayanasamy, 2009). Participation is a dynamic process where two individuals exchange their ideas, thoughts, and knowledge for making proper decisions and solving the problems faced by local people. It is a process of empowering individuals that helps people make proper analysis, make their own decisions, and develop their confidence. It is successfully used in increasing self-help activities of local people and for proper implementation and evaluation of large scale projects. FAO has reported that through participatory programmes it is very easy to mobilise local knowledge for self-reliant development, reducing government assistance costs. The decisions are taken democratically as the marginalised groups are also involved in decision making. Hence, participatory approaches act as a powerful tool for sustainable development, empowering local people and shaping their own future.

What is participation?

Active involvement of different communities, individuals and stakeholders in the development of rural sectors, different communities and agricultural processes is called participation. Participation is important in involving people in decision making, planning, implementation, monitoring and evaluation. Participation is a two way communication that helps to exchange ideas and knowledge between two individuals. It empowers different individuals and communities by providing them with different skills, knowledge and resources that are important to build their self-confidence and self-reliance.

Participatory Extension provides a framework that allows an extension agent to talk to rural people about their problems at the grassroot levels and take out some solutions (Chambers, 1992). Community participation is an important aspect where the client or beneficiary group executes the direction for implementing a particular project to increase their well being in different aspects such as personal income, growth and self-reliance.

Advantages of participatory approaches

- **Efficiency:** Participation is important as it ensures effective utilization of the available resources in a particular area. Local people are involved and take different responsibilities, resulting in resources being used more efficiently and the project becoming more cost-effective.
- **Effectiveness:** One of the major reasons for a project's failure is local people's lack of involvement in the project. Participation of local people increases effectiveness of a particular project.

- **Self-reliance:** Many governmental programmes are being undertaken in India for people so that they will have solutions for the problems faced by the individuals. The active involvement of people not only reduces dependence but also increases self-confidence and self-reliance. With the involvement of people in decision making and continuously monitoring and evaluating a programme the resources are used properly.
- **Coverage:** Developmental programs always aim to improve the society's weaker sections. Participation of people helps the target groups to benefit, and effective use of resource allocation helps wider society benefit.
- **Sustainability:** People's participation and involvement in a particular program and use of local resources leads to the development of ownership in that individual, as a result of which resources are used effectively or efficiently. Even though there may be a situation where fund allocation is ceased, they still manage funds due to their ownership of a particular resource.

(Chambers, 1994)

Steps in participatory approach

- Rapport Building: It is an interpersonal or social skill that includes establishing harmonious relations between individuals or groups of the community. The main aim of rapport building is to create mutual understanding and trust among the individuals.
- DIY (Do it yourself): The team members should make themselves involved along with the villagers in various types of activities such as sowing, transplanting of seedlings, harvesting, threshing, winnowing, etc. It not only helps in rapport building but also helps develop knowledge in practical fields and reduces complex feelings.
- Consultation with the village people: The village people and other village workers are consulted and informed before starting a programme or process. The site that is being selected should be accessible for easy supervision.
- Secondary Data Collection: Secondary data is collected from various sources such as leaflets, newspapers, reports, and library sources by NGOs, researchers, documentaries, etc. Secondary data should always be gathered before going to the field for actual diagnosis.
- Training of the trainers: It is very important for the proper training of a trainer to improve different skills and knowledge to train other extension workers and village people to ensure the sustainability of resources.
- Diagnosis: It means analysing and examining a particular situation in an area. Regarding participatory approaches, it takes care of problems, their causes and information regarding different opportunities for the villagers.

(Cavestro, 2003)

Different participatory approaches in extension

Participatory approaches focus on problem diagnosis, empowering different communities, and facilitating farmer-led research and extension (Doyle & Krasny, 2003; Loader & Amartya, 1999). Some of these approaches are as follows:

i. **Rapid Rural Appraisal (RRA):** Rapid rural appraisal is a systematic activity conducted in rural areas by multidisciplinary teams to collect new information and hypotheses regarding rural people's lives. The multidisciplinary team uses simple methods and knowledge of local people to quickly find out and analyze hypotheses and information for developing rural people, policy formulation and implementation of projects.

ii. **Participatory Rural Appraisal (PRA):** Participatory rural appraisal is a participatory approach used for developing rural people and a community-based project which involves local people in planning and assessing developmental activities. It is a bottom-up approach which starts from the grassroot level to understand the priorities of the rural people. It emphasizes the empowerment of local people to analyze their situation as an insider and act as facilitators with outsiders.

iii. **Participatory Learning and Action (PLA):** Participatory learning and approach is a learning approach where local people participate to interact among themselves, understand each other and learn from each other. PLA is a framework for analysing and generating data where local people generate qualitative and quantitative data and different information for their own analysis. It enables locals to analyze and investigate their problems, evaluate opportunities and constraints, and make decisions to achieve their goals (Richards, 1995).

iv. **Agro Ecosystem Analysis (AEA):** It was developed in Northern Pakistan with the help of the Aga Khan Rural Support System. The interaction of socio-economic processes with ecological processes creates the agroecosystem. AEA is a project's planning phase for determining many human and natural factors for the agroecosystem.

v. **Participatory Learning Methods (PALM):** This approach was developed by MYRDA in 1989 in Bengaluru. It is a non-governmental agency that aims at the development of rural people. It uses all key rules of Rapid Rural Appraisal. The main focus of PALM is to go beyond appraisal and aims at understanding rural life at grassroots levels. With a group of 25-30 externals, research is conducted in a short time to improve local conditions and socio-economic status.

Participatory Action Research (PAR): This action research was developed in the late 1940s in the United States of America. It is an approach which is achieved through learning by doing. The investigator disseminated their knowledge to the local groups. Action researchers and local groups work together to find solutions for various problems they face.

Difference between PRA and RRA

Rapid Rural Appraisal (RRA)	**Participatory Rural Appraisal (PRA)**
It was developed in late 1970's and 80's	It was developed in late 1980's an 90's
Key resource area is local people knowledge	Key resource area is local people's analytical capabilities.
Main form of innovation is method	Main form of innovation is change of behavior and attitude
Mode of RRA is extractive	Mode of PRA is facilitating
Mode of instruments is visual	Mode of instrument is verbal
Learning is done from insiders by outsiders	Learning is done by empowerment of people
Role of outsider is investigator	Role of outsider is Initiator and catalyst
Role of insider is respondent	Role of insider is analyst
RRA model is for participatory interventions	Role of PRA mode is for interaction
Donor organization demands more for RRA	Insider demands more for PRA

Space related PRA methods

1. Social Map

The main focus of the social map is to find out the habitation patterns, social infrastructure and nature of housing, including schools, roads, drinking water, drainage water, etc. It is completely done by local people, not by experts.

The main focus area for the social map is:

- To develop a comprehensive understanding of social and physical aspects of village life.
- Information is collected regarding demographic and other aspects of each household.
- A forum is provided for discussion regarding various aspects of social life.
- It serves as a guiding instrument regarding the process of planning.
- It serves as a monitoring and evaluation tool.

2. Resource Map

It focuses on the natural resources of a particular locality, including hills, rivers, land, field, vegetation, etc. It may cover habitation as well. Experts do not do resource maps. Local people do it.

Various aspects under the resource map are as follows:

- Forest, tree species and vegetation
- Terrain, topography and slopes

- Fertility, spoil type, depth and erosion.
- Command area, tenure, land and land use patterns
- Water bodies, rivers, drainage and irrigation
- Soil and water development and watershed management
- Cropping pattern, productivity, etc.

3. Transect

It is the cross-sectional representation of different agroecological zones, and there is a comparison for several parameters such as land type, topography, ownership, soil fertility, soil type, vegetation and land usage, different problems, solutions and opportunities. A resource map focuses on natural resources, whereas a transect focuses on a cross-sectional view of several agroecological zones on different parameters. It is done after the resource map for proper triangulation.

Various aspects under transect are as follows:

- To monitor and evaluate projects and interventions.
- To check out the relevancy of planned intervention.
- To verify several issues faced during PRA exercises, particularly during different mapping types.
- Appraisal in terms of problems, status and potential of natural resources.

4. Mobility Map

It is one of the PRA methods used to find out the mobility of different groups, individuals and communities. It aims to find out where people move and for what reason. It also focuses on the perception of people for movement.

Different objectives of the mobility map are as follows:

- To understand different mobility of local people, like where they move and for what reason they move.
- It is used for planning of interventions and projects.
- It is used to evaluate interventions' impacts for several effects on mobility patterns.
- Due to the increase in awareness and gender sensitivity, it is highlighted to show the difference between mobility patterns for women and men.

Time related PRA methods

1. Time Line

It is the most commonly used PRA method. Time line is used to capture the events chronologically that local people recall. A sequential aggregate is drawn from the past events. Historical landmarks are provided for a community, an individual and an institute.

The main objectives under time line are as follows:

- It is used to learn from the community that all events are considered important in past events.
- It is used to understand different historical perspectives from the community on recent issues.
- It is used to generate several discussions on issues such as gender, education, health facilities, food security, and socio-economic conditions.
- It is used to develop rapport with different village people as discussions related to the village with the villagers will always be interesting for them.

2. Trend Analysis

It is one of the popular PRA methods used to find out temporal dimensions with respect to different changes. The local people have a good idea regarding recent scenarios, keeping in mind what things have changed eventually. It is helpful to find out all quantitative changes related to different aspects of village life, such as livestock, yield, population, areas under cultivation, rainfall, number of trees, etc.

Several objectives under trend analysis are as follows:

- To learn from the community how they have perceived change related to several aspects of life over time.
- Significant changes in the village profile are integrated.
- Various village problems are discussed regarding the increase or decrease of problems over the years.
- Different measures or interventions are discussed which has failed or worked out in the past.
- People's perception of the past or present is understood.
- After discussions, a conductive environment is produced to plan for present and future interventions.

3. Seasonal Diagram

It is also known as seasonal activity, seasonal calendar, and seasonal analysis. It is one of the popular methods under PRA that has been used over seasons and months for proper analysis. This PRA method is not done under statistics. It is done by triangulating primary and secondary data for the verification of information that is being collected.

Objectives under the seasonal diagram are as follows:

- It is used to identify food security, wage availability, diseases, heavy workload, etc.
- It is one of the important factors for project implementation as it takes care of the particular time when the project should be implemented.

- It identifies the period of stress and is also used to plan when the intervention is required.
- It analyses different patterns of livelihood over the years.

Relation related PRA methods

1. Cause effect diagram (Problem Tree)

It is called a fishbone or Ishikawa diagram and falls under linkage and flow diagram methods. It focuses on causes and effects presented visually and their interlinkage, which helps in an in-depth study of a particular topic and helps local people properly analyse it. Cause effect depicts various causes identified by local people, such as illiteracy, poverty, child labour, and health services. The advantages of a cause-effect diagram are:

- It focuses on causes rather than symptoms.
- It provides local people with the opportunity to express the complex realities.
- It depicts the visual representation of various causes and effects of problems.
- It discusses all possible solutions to a problem after knowing its cause and effects.

Objectives of the cause-effect diagram are as follows:

- It identifies the cause and effect of a particular problem or phenomenon.
- It is used to identify all possible causes of a problem to find a solution.

2. Well being ranking methods

It is also called as well being ranking or wealth analysis, which is one of the most commonly used PRA methods for ranking different communities and households on the basis of wealth, income and other criteria. Well being ranking method is a relative assessment, not an absolute assessment. Well being ranking methods is done through several methods such as

- ➢ Card sorting method
- ➢ Matrix method.
- ➢ Social mapping method.

Objectives of the well-being ranking method:

- Classification of households and groups are identified based on well being in certain criteria such as wealth, income, etc.
- Target groups are identified and checked to see if the programmes fulfil proper needs and requirements.
- Inter-group and inter-household disparities are studied and understood in how local people view it.

- The impact of interventions is studied on different households, and a baseline is developed for evaluation.

3. Venn Diagram

It is one of the PRA methods used to study institutional relationships, also called institutional diagrams. It is also called as chapatti diagram in Hindi. It is assumed that the bigger the circle, the more important an institute or individual will be. Distance between two circles is the degree of contact between two institutes or individuals. Overlapping circles indicate that the interaction is stronger.

Objectives of the Venn diagram are as follows:

- It is used to understand people's perceptions of individuals, local institutes and programmes.
- It provides valuable insights to analyse the decision-making system.
- The importance of a particular programme or service can be studied under the Venn diagram.

Conclusion

It is thus concluded that the participatory approach in extension is an effective and powerful way to empower several communities to increase the effect of extension programmes. It is also used for the promotion of sustainable development. Local leaders, experts, and farmers are actively involved in making decisions to build trust and fulfil the needs and requirements of targeted people. It is used for bridging the gaps between researchers, extension experts and local people, leading to successful intervention of a particular project or programme. Gender equity is supported by the participatory approach that takes care of traditionally marginalized groups with a voice in the developmental process. Due to active participation, the participatory approach has an important role in sustainable agriculture and rural development. Despite many challenges, participatory approaches positively impact motivating people and improving the livelihood of rural communities. Hence, participatory approaches play a pivotal role in achieving goals and ultimately create positive impacts on rural communities, agriculture and the environment.

References

Cavestro, L. (2003). PRA-participatory rural appraisal concepts methodologies and techniques. Padova University. Padova PD. Italia.

Chambers, R. (1992). Rural apprasial: rapid, relaxed and participatory. Institute of Development Studies (UK).

Chambers, R. (1994). Participatory rural appraisal (PRA): Challenges, potentials and paradigm. World development, 22(10), 1437-1454.

Doyle, R., & Krasny, M. (2003). Participatory rural appraisal as an approach to environmental education in urban community gardens. Environmental Education Research, 9(1), 91-115.

Loader, R., & Amartya, L. (1999). Participatory rural appraisal: extending the research methods base. Agricultural systems, 62(2), 73-85.

Narayanasamy, N. (2009). Participatory rural appraisal: Principles, methods and application. SAGE Publications India.

Richards, P. (1995). Participatory rural appraisal: a quick and dirty critique. PLA notes, 13-16.

13

Empowering Rural Progress: Schemes for Upliftment of Rural Development

Wandahun Lynshiang*[1] and Bhaskar Pal[2] and Vicky Yadav[3]

[1]*Department of Extension Education and Communication Management College of Community Science, RPCUA, Pusa, Samastipur-848125, Bihar*
[2,3]*Department of Extension Education, PG College of Agriculture, RPCAU Pusa, Samastipur-848125, Bihar*
**Corresponding author E-mail:* wandahunlynshiang404@gmail.com

Abstract

The rural development upliftment project is a comprehensive initiative aimed at supporting long-term growth and prosperity in rural communities. This comprehensive approach includes crucial components such as infrastructural development, access to education and healthcare, agricultural breakthroughs, skill training, and financial inclusion. By addressing these essential pillars, the initiative aims to empower rural communities, eliminate poverty, and improve their general quality of life, so contributing to the nation's inclusive progress. A strategy for the upliftment of rural development is often a government or nonprofit initiative aiming at improving the living circumstances, infrastructure, and overall well-being of rural communities. Such initiatives frequently cover many facets of rural development, such as agriculture, education, healthcare, infrastructure, and economic prospects. Rural development schemes serve a critical role in bridging the urban-rural gap and supporting equitable growth. These programs are critical for empowering rural communities, improving their socioeconomic prospects, and ultimately contributing to a country's overall development. Governments and organizations may harness the immense potential of rural areas by investing in them, creating a more equal and prosperous future for all citizens.

Keywords: *Development, India, Rural, Scheme, Well being.*

Introduction

Terms total development of rural areas with the purpose of increasing the quality of life for rural people is referred to as rural development. It is a broad and diverse concept that encourages the expansion of agricultural and allied activities, communities, cottage industries and crafts, community services and facilities, and total human resources in rural areas. As a strategy, rural development is the result of interactions between many physical, technological, economic, and societal elements. Its purpose is to improve the economic and social well-being of a particular rural demographic group. Rural development is a multifaceted concept that covers various aspects of development in order to improve rural people's quality of life (Ilbery, 2008). Agriculture, village industries, crafts, community services, and rural human resource development all fall under this category. This holistic approach examines the interplay of physical, technological, economic, and sociocultural variables as a strategic means of attaining its goals (Hagen, 1962). Finally, rural development seeks to improve the economic and social well-being of rural communities by focusing on specific groups within them.

Rural development is a technique for assisting a certain group of people, the poor rural population, to obtain more of what they desire and require for themselves and their children. It means aiding the poorest among those seeking a living in rural areas to demand and control a bigger portion of the benefits of rural development. Rural development as a process improves the quality of life for rural populations, particularly the poor, through time. It is a method of enabling impoverished rural residents to better their own and their children's lives by getting access to the resources and opportunities they want. This technique aims to assist the most vulnerable rural residents in asserting their rights and influencing the outcomes of rural development activities. Finally, rural development attempts to encourage long-term improvements in the well-being of rural communities, particularly underprivileged ones.

Concepts of rural development

According to Seers (1969), growth requires more than just rising earnings. It is a process of expanding people's true liberty. Thus, progress means lowering significant sources of unfreedom, such as poverty and dictatorship, restricted economic prospects, as well as systematic social misery, neglect of public infrastructure, and intolerance or over-activity by oppressive authorities.

Objectives of upliftment of rural development

Uplifting rural development is a vital component of promoting equitable economic growth and improving rural populations' quality of life (Hazell and Wood, 2008). To accomplish this goal, many objectives and strategies are employed. The following are some common goals for improving rural development:

- Poverty alleviation: reduce and eventually eliminate poverty in rural areas by creating income-generating opportunities for the rural population.

- Infrastructure development: improve basic infrastructure like roads, electricity, clean water supply, and sanitation facilities to enhance living standards and accessibility.
- Agricultural modernization: promote modern farming techniques, provide access to better seeds, fertilizers, and machinery to increase agricultural productivity and income.
- Skill development: offer training and skill development programs to rural residents, enabling them to access better employment opportunities, both within and outside agriculture.
- Healthcare services: ensure access to quality healthcare services, including clinics and hospitals, to improve the overall health and well-being of rural communities.
- Education and literacy: establish schools and educational institutions to promote literacy and provide children with a better chance at a brighter future.
- Rural employment generation: promote small and medium-sized enterprises (SMEs) and cottage industries in rural areas to create job opportunities and reduce urban migration.
- Access to credit and financial inclusion: facilitate easy access to credit and banking services to help rural entrepreneurs and farmers invest in their businesses and save money.
- Women empowerment: implement programs that empower women in rural areas through education, skill training, and opportunities for income generation.
- Natural resource management: promote sustainable practices in agriculture and natural resource management to protect the environment and ensure long-term livelihoods.
- Community development: foster a sense of community involvement and ownership in development projects, ensuring that local communities have a say in their development priorities.
- Promotion of cooperatives: encourage the formation of agricultural cooperatives and self-help groups to pool resources, share knowledge, and collectively benefit from economies of scale.

Schemes for upliftment of rural development

Schemes	Year of establishment
Sarva Shiksha Abhiyan	2001-2002
Mahatma Gandhi National Rural Employment Guarantee Act	2006
Rashtriya Krishi Vikas Yojana	2007
Deendayal Antyodaya Yojana	2011
Swachh Bharat Abhiyan	2014
Indira Awaas Yojana	2015
Pradhan Mantri Suraksha Bima Yojana	2015
Ayushman Bharat	2018
Pradhan Mantri Kisan Samman Nidhi	2018-2019

Source: Jagranjosh, 2023; Ministry of Rural Development Government of India, 2023

1. Sarva Shiksha Abhiyan (SSA)

The Sarva Shiksha Abhiyan (SSA) is a government of India program that aspires to attain universal primary education within a set time limit. The program was launched in 2001 as a flagship initiative of the Indian government's Ministry of Human Resource Development. Its major purpose is to provide free and compulsory education to all children between the ages of six and fourteen, with a particular emphasis on children from marginalized and underprivileged communities. The Sarva Shiksha Abhiyan has played a critical role in increasing access to primary and upper-primary education throughout India, resulting in a significant reduction in the number of out-of-school children. It has been crucial in improving overall educational quality in government schools and promoting inclusive education techniques.

Objectives of the Sarva Shiksha Abhiyan include

- Universal enrollment: ensuring that every child in the target age group is enrolled in a school and has access to quality education.
- Infrastructure development: building and upgrading school infrastructure, such as classrooms, libraries, and sanitation facilities, to create a conducive learning environment.
- Quality improvement: enhancing the quality of education by providing necessary teaching and learning materials, teacher training, and academic support.
- Financial support: providing financial assistance to states and union territories to implement SSA effectively.
- Special emphasis on girls and marginalized groups: tackling gender and social gaps in education by encouraging girls and children from disadvantaged families to enroll and remain in school.

- Alternative and innovative education: implementing innovative strategies to cater to the needs of out-of-school children and those with special needs.

2. Mahatma Gandhi National Rural Employment Guarantee Act (MGNREGA)

The Legislative Assembly approved it on August 25th, 2005, and the system went live in 2006. The 'National Rural Employment Guarantee Act' was the initial name. The Act was enacted with the intention of enhancing the purchasing power of persons in rural India who are semi- or unskilled, whether poor or not. The programs allow adult members of any rural household who want to do public work related to unskilled manual labor to do so at the statutory minimum wage of Rs 100 per day, with a legal guarantee of 100 days every fiscal year. As a result, MGNREGA aims to achieve two objectives: rural development and employment creation.

- Financial security: the fundamental goal of MGNREGA is to provide rural households with financial security by providing employment opportunities. This ensures that they have a minimal level of income even when agricultural and work conditions are difficult.
- MGNREGA intends to develop sustainable job possibilities in rural communities, particularly during lean agricultural seasons, minimizing seasonal migration to metropolitan areas in pursuit of work.
- Rural infrastructure development: the program focuses on the creation of productive and long-lasting rural infrastructure such as roads, irrigation canals, and water conservation structures, among other things. These assets contribute to the general quality of life in rural regions while also increasing agricultural output.

3. Rashtriya Krishi Vikas Yojana (RKVY)

The Rashtriya Krishi Vikas Yojana is a government program in India that encourages agricultural development. The RKVY debuted in 2007-08.The RKVY provides financial assistance to Indian states and union territories for a variety of agricultural and allied sector projects and activities. These programs may include infrastructure upgrades such as irrigation, rural roads, and storage facilities, as well as encouragement of activities such as crop diversification, organic farming, and agribusiness expansion. The RKVY has been instrumental in encouraging agricultural growth and rural development in India by providing financial and technical assistance to numerous states for agricultural development programs.

The primary objectives of the RKVY include:

- Advancing the overall growth of agriculture and allied sectors through the implementation of modern technology and best agricultural practices.
- Ensure equal distribution of resources and benefits to all farmers in India's various regions.

- Improving agriculture's production and sustainability.
- Improvements to agricultural infrastructure and support services.
- In agriculture, encouraging public-private cooperation.
- Increasing agricultural output value addition and post-harvest management.
- Providing job prospects in remote communities.

4. Deendayal Antyodaya Yojana (DAY)

Deendayal Antyodaya Yojana (DAY) is an Indian government flagship program aiming at reducing poverty and empowering the poorest of the poor in rural and urban parts of the country. The scheme was started in 2016 by the Ministry of Rural Development and is named after Pandit Deendayal Upadhyaya, a significant leader and philosopher of the Bharatiya Janata Sangh (a precursor of the Bharatiya Janata Party).

➢ The Deendayal Antyodaya Yojana has two main components:

I.National Rural Livelihoods Mission (NRLM):

- The NRLM, also known as Aajeevika, promotes self-employment and organizes rural poor into Self-Help Groups (SHGs).
- Its key goals include promoting social and economic empowerment to women through SHGs, as well as creating sustainable lives for rural communities.
- The NRLM seeks to cover all rural districts in India and reach millions of poor households.

II.National Urban Livelihoods Mission (NULM):

- The NULM strives to eliminate poverty and vulnerability in urban areas by expanding the urban poor's livelihood options.
- It focuses on skill training and placement, self-employment programs, and increasing access to financial services in urban disadvantaged neighborhoods.
- NULM also prioritizes shelter for the urban homeless and social security for vulnerable city dwellers.

The key goals and objectives of the Deendayal Antyodaya Yojana include:Objectives of the National Rural Livelihood Mission include:

- Self-Help Groups (SHGs): The NRLM encourages the formation and expansion of SHGs, which are small groups of rural women who meet to save money, seek credit, and engage in income-generating activities. SHGs are essential for improving financial inclusion and empowering women.
- Livelihood enhancement: enhancing the income and livelihood prospects of the poor by providing training, skill development, and support for self-employment and wage employment opportunities.

- Housing and social security: providing housing and social security measures for the urban homeless and vulnerable populations.

5. Swachh Bharat Abhiyan (SBA)

Swachh Bharat Abhiyan, often known as the Clean India Mission, is a nationwide cleanliness effort launched by the Indian government on 2nd October, 2014. It is one of Prime Minister Narendra Modi's flagship initiatives, with the goal of making India clean and defecation-free by October 2, 2019, to coincide with Mahatma Gandhi's 150th birth anniversary, who advocated cleanliness and sanitation as essential aspects of a healthy and dignified life. The Swachh Bharat Abhiyan has made significant progress in increasing sanitation and cleanliness across India. While the campaign's original objective of 100 percent cleaning by 2019 was lofty, it did raise awareness of hygiene and sanitation requirements.

Objectives of the Swachh Bharat Abhiyan include:

- Construction of toilets: one of the primary focuses of the campaign was to build toilets in rural and urban areas to eliminate open defecation. This involved constructing millions of toilets across the country.
- Behavioral change: the campaign aimed to bring about a behavioral change in people's attitudes towards cleanliness and sanitation. It included educational and awareness programs to promote hygiene and cleanliness practices.
- Solid waste management: proper disposal of solid waste was emphasized to keep streets, neighborhoods, and public places clean. This included waste segregation and recycling initiatives.
- The program aims to declare all villages, towns, and cities in India as Open Defecation-Free (ODF) by encouraging the construction and usage of toilets.
- Cleanliness rankings: the government introduced a Swachh Survekshan (Cleanliness Survey) to assess and rank cities and towns based on their cleanliness and sanitation efforts.
- Sanitation coverage: expanding sanitation coverage in schools, colleges and other public institutions was another important aspect of the campaign.

6. Indira Awaas Yojana (IAY)

The Indira Awaas Yojana (IAY) was a government of India project that began in 1985-86 as part of the Rural Landless Employment Guarantee Programme (RLEGP) and later evolved into a sub-scheme of the Jawahar Rozgar Yojana (JRY). However, on 1st January, 1996, it became an autonomous system. On April 1st, 2016, the IAY was revamped and renamed the Pradhan Mantri Awaas Yojana Gramin (PMAY-G), indicating the government's commitment to addressing rural housing needs and empowering rural communities through housing assistance (Gohil and Gandhi, 2019).

- The main purpose of IAY was to provide free housing units to specific target groups in rural areas, including as Scheduled Castes, Scheduled Tribes, free-bonded laborers below the poverty line, and relatives of ex-servicemen of the armed and paramilitary forces who were killed in action. Furthermore, 3 percent of the dwellings were reserved for disabled persons living in rural areas who are poor.
- The scope of IAY was gradually enlarged to cover non-scheduled caste and non-scheduled tribal rural poor, with a 40 percent cap on payments made to this group. The Central Government contributed 75 percent of IAY money, while the State Government contributed 25 percent.
- The effort aimed to play a significant role in rural development by providing financial assistance to fewer rurally disadvantaged individuals and families. It was a crucial tool for improving the living conditions of rural India's disadvantaged and vulnerable groups.

7. Pradhan Mantri Suraksha Bima Yojana (PMSBY)

India Prime Minister Mr. Narendra Modi, has emphasized financial inclusion and launched many public welfare projects for inclusive finance through insurance since 2014. The Pradhan Mantri Suraksha Bima Yojana (PMSBY), which was launched on 9th May, 2015, is a significant strategy aimed at building a universal social insurance system, with a focus on the poor and vulnerable (Kumar, et al., 2020). PMSBY is designed to offer low-cost accident insurance coverage to high-risk individuals who travel frequently, such as mechanics, laborers and truck drivers. It covers accidental death, permanent disability, and partial disability and has one-year insurance terms that are renewable annually.

Objectives of the PMSBY are:

- To raise public awareness of the PMSBY plan and its components.
- To assess the current state of the PMSBY Scheme.
- To investigate the enrolment and claim distribution ratio of the PMSBY plan by state or union territory.
- To make recommendations for improving the PMSBY program's implementation.

8. Ayushman Bharat

Ayushman Bharat, also known as the Pradhan Mantri Jan Arogya Yojana (PM-JAY), is a significant healthcare scheme implemented by the Government of India in September 2018. This ambitious program aims to provide financial security and healthcare access to India's underprivileged and economically disadvantaged households. Ayushman Bharat is one of the world's largest government-funded healthcare insurance programmes, with the purpose of decreasing financial burdens on families during medical emergencies while also enhancing access and outcomes. It has made a significant contribution to India's efforts to achieve universal health coverage and improve the general health and well-being of its citizens.

➢ Ayushman Bharat has two main components:

I.Pradhan Mantri Jan Arogya Yojana (PM-JAY):

- PM-JAY provides health insurance coverage to about 10 crores (100 million) economically disadvantaged families in India, with roughly 50 crores (500 million) beneficiaries. Under this scheme, qualified families are entitled to financial assistance for hospitalization and medical treatment up to Rs. 5 lakhs (5,00,000 Indian Rupees) per family each year.
- PM-JAY covers a wide range of medical treatments and procedures, including operations, diagnostic testing, hospitalization, and other important healthcare services.
- The scheme primarily targets individuals and families living in poverty, as well as those defined as "deprived rural families" and "deprived urban families" using data from the Socio-Economic Caste Census (SECC).

II.Health and Wellness Centers (HWCs):

- Ayushman Bharat also aspires to strengthen and transform India's basic healthcare system by establishing Health and Wellness Centers (HWCs).
- These HWCs are intended to bring comprehensive healthcare services closer to people's homes, including preventative, promotional, and basic curative treatment. They concentrate on non-communicable diseases, maternity and child health, and other issues.
- The goal is to establish a network of 1.5 lakh (150,000) HWCs across the country to ensure that everyone have access to high-quality primary care.

9. Pradhan Mantri Kisan Samman Nidhi (PM-KISAN)

The Pradhan Mantri Kisan Samman Nidhi (PM-KISAN) is a government-funded agricultural assistance program in India. The Government of India formed it in February 2019 and it is administered by the Ministry of Agriculture and Farmers Welfare. The primary purpose of PM-KISAN is to assist small and marginal farmers with agricultural expenses and to improve their livelihoods.

Objectives of the PM-KISAN scheme are:

- Poverty alleviation: the scheme tries to eliminate rural poverty by giving financial assistance to farmers, particularly those with little landholding. The income support is meant to help farmers and their families improve their financial situation.
- Increasing agricultural production: the initiative indirectly tries to increase agricultural productivity by providing farmers with the financial resources

they require to invest in improved seeds, fertilizers, and other inputs. This can result in higher crop yields and better agricultural practices generally.

- Strengthening the rural economy: by increasing farmer income, PMKISAN contributes to the general strengthening of the rural economy. Increased rural income can lead to increased rural economic activity, benefiting not only farmers but also rural enterprises and communities.

Conclusion

Rural development schemes are efforts and policies that try to improve rural people's quality of life, economic prospects, and infrastructure. Governments, non-governmental organizations (NGOs), and international organizations often execute these projects to meet the unique issues that rural people experience. The precise goals and target areas of rural development programs or schemes may differ from region to region, but the general objectives remain. Rural development projects are frequently tailored to the unique demands and problems of each location. Government finances, international aid, grants, and public-private partnerships may all be used to fund them. A comprehensive and inclusive approach comprising collaboration among numerous stakeholders such as local communities, government agencies, non-governmental organizations (NGOs), and the corporate sector is typically required for successful rural development.

References

Gohil, J., and Gandhi, Z. H. (2019). Pradhan Mantri Awas Yojana (PMAY) Scheme—an emerging prospect of affordable housing in India. International Research Journal of Engineering and Technology, 6(12), 2546-2550.

Hagen, E. E. (1962). "On the Theory of Social Change: How Economic Growth Begins." Homewood, IL: Dorsey Press.

Hazell, P., and Wood, S. (2008). Drivers of Change in Global Agriculture. Philosophical Transactions of the Royal Society B: Biological Sciences, 363(1491), 495-515.

Ilbery, B. (2008). "Changing ruralities and the challenge for rural development." In The Oxford Handbook of Rural Planning Oxford University Press, pp:25-47.

Jagranjosh. (2023). Lists of Various Rural Development Schemes in India. https://www.jagranjosh.com/general-knowledge/list-of-various-rural-development-schemes-in-india-1480939663-1

Kumar, M. S., Pandian, P. S., and Nagarajan, N. R. (2020). A study on social insurance schemes in India with special reference to Pradhan Mantri Suraksha Bima Yojana scheme. ICTACT Journal on Management Studies, 6(2), 1233-1237.

Ministry of Rural Development Government of India. (2023). Schemes/ Programmes. https://rural.nic.in/en/scheme-websites

Seers, D. (1969). "What Are We Trying to Measure?" Journal of Development Studies,5(3), 5-15.

14

Communication Methods and Theories

Saloni Kumari[1] and Vicky Yadav*[2]

[1]Research scholar, Dept.of Extension Education and Communication Management, College of Community Science
[2]Research scholar, Department of Agricultural Extension Education, PG College of Agriculture, Dr. Rajendra Prasad Central Agricultural University, Pusa, Samastipur-843125, Bihar.
**Corresponding author E-mail: Vickyspecial01@gmail.com*

"Communication- the human connection- is the key to personal and career success"

(Paul J. Meyer- Motivator)

Abstract

Communication is a fundamental aspect of human interaction, essential for understanding and collaboration. Effective communication plays a pivotal role in personal relationships, professional success, and societal cohesion. Clear and concise communication is vital in conveying intentions, resolving conflicts, and fostering connections. Communication is the cornerstone of human connection and a linchpin for social progress. Cultural context also emerges as a critical factor. The chapter provides a comprehensive exploration of communication, encompassing its meaning, core elements, diverse perspectives, methods, and key theories. It delves into the fundamental concept of communication as the exchange of information, thoughts, and emotions. It dissects the essential elements of sender, message, medium, receiver, and feedback, elucidating their interplay. It offers an insightful analysis of various perspectives, including mechanistic, psychological, social, systemic, and critical viewpoints. Moreover, it delves into pivotal communication theories such as normative, authoritarian, free press, social responsibility, developmental, and democratic theories, providing a robust framework for understanding this crucial aspect of human interaction.

Keywords: *Communicator, Communication Models, Message, Methods, Theories.*

Introduction

The term 'communication' originates from the Latin word 'Communis', means the act of imparting, participating, and transmitting or the notion of making something common or sharing it among individuals. Communication encompasses the multifaceted process of sharing and exchanging ideas, emotions, information, and experiences between two or more individuals. It embodies the act of transmitting thoughts or expressions, serving as a vital conduit for the flow of knowledge and understanding. Essentially, it is aligning a sender and receiver for a specific message or series of messages. Leagans (1961) states that communication is the process by which two or more people exchange ideas, facts, feelings or impressions in ways that each gains a common understanding of the meaning, intent and use of messages. Effective communication is not just about issuing orders; it is about fostering comprehension (Suresh, 2003). It beyond imparting knowledge to enabling a profound understands (TNAU, 2012). Future progress hinges on advanced technology and adept communication of wrong messages, their timing, and audience hinder progress. What is vital is more individuals voicing the right messages to the right audience at the right moment, in the right manner. This is the key points of good communication.

Table 1: Definition of Communication

S.No.	Definition	Source
1.	Communication is the discriminatory response of an organism to stimulus.	(Stevens,1942)
2.	Communication is all the procedure by which one mind can affect another.	(Shannan and Weaver,1949)
3.	Communication is the mutual interchange of ideas by any effective means.	(Wilbur schramm,1955)
4.	Communication is the process of transferring messages from a source to a receiver.	(Rogers and Shoemaker, 1971).
5.	Communication is a purposeful process involving sources, messages, channels, and receivers.	(Andersch et al, 1969)
6.	Communication is the process of understanding and sharing meaning.	(Pearson and Nelson, 2000)
7.	Communication is a basic human activity that enables us to connect.	(Langley, 2006)
8.	Communication is how people share their ideas, information, opinions and feelings with others.it is a two-way process and therefore requires complete understanding among the communicators.	(Senapathi, 2010)

(***Source***: ICAR e-Cource, 2012; TNAU, 2012)

Table 2: Elements of Communication

S.No.	Elements	Deliberation	Example
1.	Communicator	Originators of message	In farming and rural development, a person like an extension worker or a communicator starts the talk. It's not just them or the radio giving out information; it actually comes from places like research institutes and agricultural universities. The extension worker learns from this research and tells the farmers what they need to know, sometimes using helpful tools like pictures or videos. They also listen to the farmers' thoughts and problems and share it back with the researchers to find solutions
2.	Message	The content or subject matter	When we talk about farming advice or technology, that's what we call the subject. A good message should be clear about what needs to be done, how to do it, when to do it, and what to expect as a result. otherwise, it's just like background noise. Messages that are important, interesting, helpful, profitable, trustworthy (based on the latest and best research), and not too overwhelming or too simple are more likely to get people motivated.
3.	Channel	The Medium	Farm & Home Visit by Subject matter specialist of Krishi Vigyan Kendra to show demonstrate the new seed of paddy in farmers' field.
4.	Treatment of message	The way a message is handled for the information gets across the audience	To use of local language in the field demonstration for the better understanding of framers and rural people.
5.	Audience	Receiver of message	Field crop farmers, fruit growing farmers, dairy men, poultry keepers, fish farmers, homemakers etc.
6.	Audience responses	Feedback	After using the new developed seed which gave to farmers by extortioner they gave feedback about the per capita yield.

Communication Models

1. Aristotle's model (before 300 BC)

- First basic Persuasive Communication Model

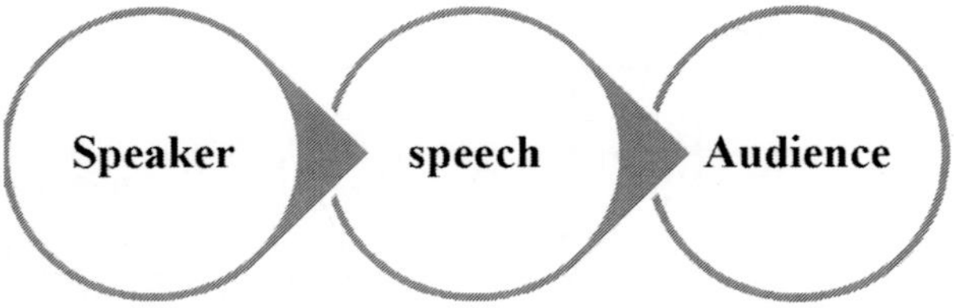

Fig. 1: Aristotle's model

2. Lasswell's model (1948)

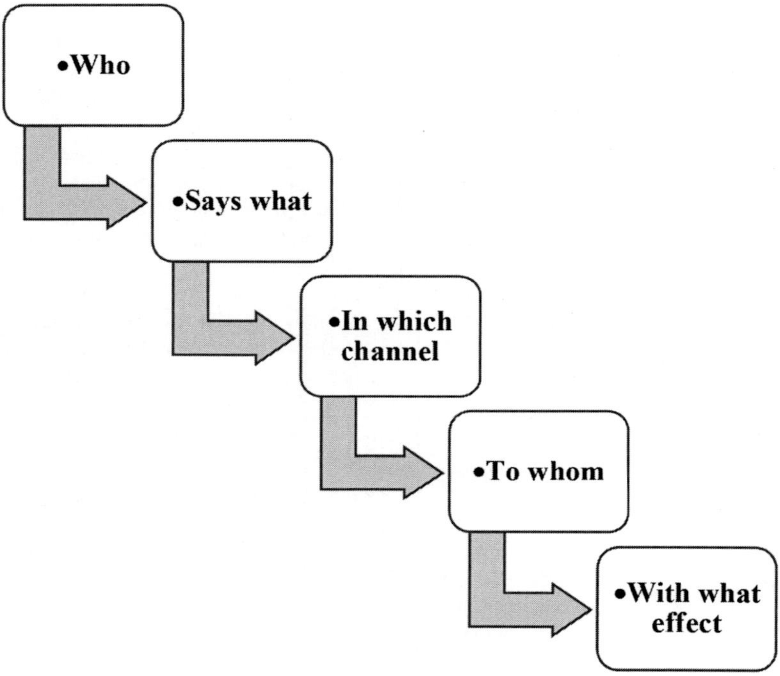

Fig. 2: Lasswell's model

3. The Shanon- Weaver model (1949)

- Mathematical theory of communication
- Information theory
- Best suited for broadcasting of farm information
- Noise is important element

Fig. 3: Shanon- Weaver model

4. Berlo model (1960)

- The Process of mass Communication Book Written by Berlo
- Types of Skills Pointed out by Berlo are, Encoder and Decoder

Fig. 4: Berlo model

5. Schramm model (1961)

- Destination is important component

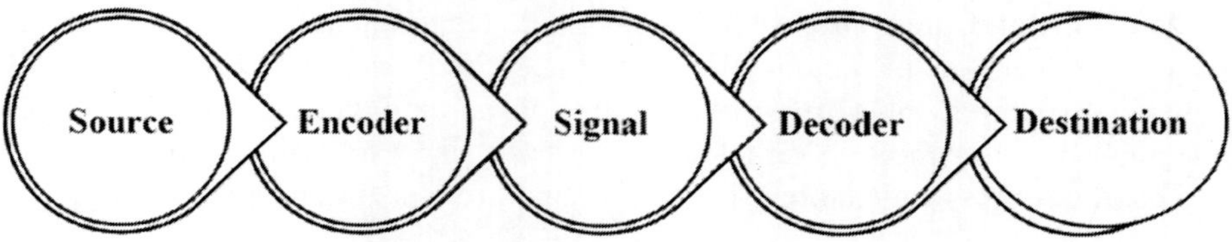

Fig. 5: Schramm model

6. Leagans model (1963)

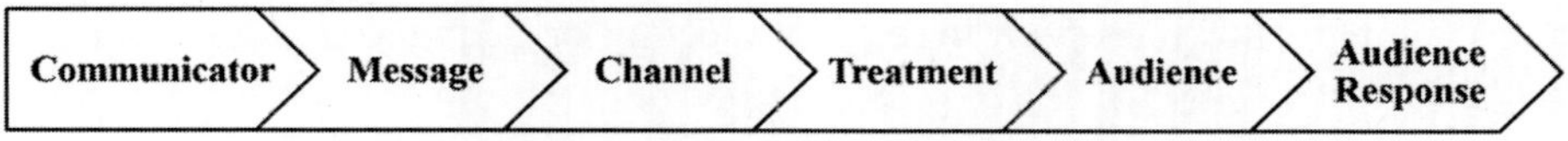

Fig. 6: Leagans Model

7. Helical model (Frank Dance, 1967)

Cyclical,

Contextual (Influenced by time and experience),

Continuous,

Non-Repetitive, and.

Accumulative (getting increasingly more complex and 'knowledgeable')

8. Rogers and Shoemaker model (1971)

1. Source
2. Message
3. Channel
4. Receiver
5. Effects

Methods of Communication

A. **Verbal and non- Verbal Communication**: Communication is categorized into verbal and non-verbal based on the mode of expression and the channels used to convey a message.

1. Verbal Communication: Verbal communication involves the use of spoken or written words to convey a message. It relies on language and grammar to communicate ideas, emotions, information, and instructions. It primarily occurs through spoken language, written words, or sign language for those with hearing impairments.

Examples: Conversations, speeches, presentations, letters, emails, text messages, etc.

- Oral – like face-to-face communication, lectures, phone calls, seminars, etc.
- Written- Letters, Email, SMS, etc.

2. Non-Verbal Communication: Non-verbal communication encompasses all forms of expression that do not involve words. It relies on facial expressions, body language, gestures, tone of voice, and other visual and auditory cues to convey meaning. It occurs through visual cues (such as facial expressions, body posture, and hand gestures) and auditory cues (like tone of voice, pitch, and volume).

- **Chronemics:** Itrefers to how people use and perceive time in communication. It includes things like being punctual, how long you spend with someone, and the pace at which you speak.

For example, showing up on time for a meeting or taking your time to have a leisurely conversation.

- **Proxemics:** It is all about how people use and understand personal space in communication. It's like an invisible bubble around each person.

For example, if someone stands really close to you when talking, you might feel uncomfortable because they're in your "personal space.

- **Kinesics:** Kinesics is the study of body language and facial expressions in communication.

Example: a smile, a wave, or even a nod can all be forms of kinesics.

- **Vocalis:** Itis all about how we use our voice when we communicate. It includes things like tone, pitch, volume, and even pauses.

For example, when someone speaks loudly and with excitement, that's using vocalics to convey their emotions.

- **Haptics:** Haptics is the study of touch in communication. It's about how we use physical contact, like hugs, handshakes, or pats on the back, to convey emotions, support, or information.

Example: a comforting touch on the shoulder can express empathy, while a firm handshake can indicate confidence.

- **Artifacts:** Artifacts in communication refer to the objects or possessions that people use to express their identity, personality, or values. These can include clothing, accessories, symbols, and even the arrangement of personal items in one's environment. For example, a person's choice of clothing style or the decorations in their living space can communicate information about their cultural background, interests, or social status.

B. Formal and informal communication:

1. Formal Communication: Formal Communication is a structured way of exchanging information within an organization, guided by established protocols and channels. This type of communication adheres to the hierarchical structure set by the organization itself. Examples include reports, directives, memos, and official announcements. It is also known as official communication.

- **Vertical communication**: it involves information exchange between different levels of an organization.
➢ **Top-down**-It can flow from higher-ups to subordinates
➢ **Bottom- down-** from subordinates to higher-ups
- **Lateral or Horizontal:** This type of communication occurs between peers working in different departments at the same organizational level. Example- Communication between a sales manager and human resource manager.
- **Diagonal or crosswise**: Inter-departmental communication occurs between employees in different departments and levels. Example- Communication takes between the salesman and manufacturing manager.

2. Informal communication: It is a multidimensional approach to communication and flows freely in the organisation without any restraint of predetermined channels for transmission of information. it is known as grapevine communication. Examples: casual discussion between team members, gossips, tea clubs etc.

❖ Single Strand Chain: Information passes from one person to the next in a linear sequence.
❖ Cluster Chain: One person shares information with a group of friends, spreading it within their own circles.
❖ Probability Chain: Information is randomly shared with selected individuals who continue the process.
❖ Gossip Chain: One person shares information with a group, who then pass it on to others, leading to widespread dissemination (Board Infinity, 2023).

C. Intra and Interpersonal Communication:

1. Intrapersonal communication: it happens internally within oneself, occurring in our thoughts i.e., sender and receiver is the same. self-concept i.e., beliefs, values and attitudes are the basis for interpersonal communication.

- **Internal discourse-** Internal discourse refers to the ongoing inner conversation and thinking process that occurs within an individual's mind.

Example: When making a decision, a person may weigh the pros and cons in their mind, considering various options and possible outcomes, engaging in internal discourse before arriving at a conclusion.

- **Solo vocal communication**- Solo vocal communication is when an individual communicates using their voice without any direct interaction from others.

Example: Singing alone in a room, expressing emotions and thoughts through song without any audience or feedback.

- **Solo written communication**- Solo written communication involves an individual conveying a message through written means without immediate feedback or interaction.

Example: Writing a personal journal entry, where one expresses their thoughts, emotions, and experiences on paper without expecting immediate response from others.

2. Interpersonal communication: it is the exchange between individuals whose lives interact and affect each other.

- **Dyadic Communication**: Dyadic communication is a conversation between two people.

Example: A one-on-one conversation between friends discussing weekend plans.

- **Group Communication**: it involves three or more persons. Example: Panchayat meetings, United Nations Assembly etc.
- **Public Communication**: Public communication involves one person addressing a large audience.

Example: A politician giving a speech to a crowd at a campaign rally.

Theories of Communication

Communication theory, proposed by S. F. Scudder in 1980, says all living things communicate in different ways. For example, plants show they need care through changes in their appearance. Animals use sounds and movements to express themselves. Communication is vital to our lives, influencing everything we do and how we connect with others. Theories help us understand this better and build on what we already know. It's not just about one theory, but a whole body of knowledge about communication. Our daily lives are shaped by how we talk to others and even by messages from people we have never met. They can be from anywhere, anytime, and greatly impact us.

Perspective of Communication and communication theory: Communication theories examines how people exchange information and understand each other, offering insights into the various ways messages are sent, received, and interpreted in different contexts. It helps us understand the complex dynamics of human interaction and the role of communication in shaping societies and relationships.

Table 3: Perspective viewpoints of Communication and communication theory

S. No.	Domain	Deliberation
1.	Mechanistic	Communication is sending information from sender to receiver."
2.	Psychological	Communication encompasses the sender's thoughts, feelings, and the recipient's reactions and emotions, extending beyond mere information transmission.
3.	Social	Communication is a product of sender-receiver interaction, emphasizing the importance of both content and delivery.
4.	Systemic	Communication generates a unique message as individuals interpret, reinterpret, and draw their conclusions from it.
5.	Critical	Communication serves as a means for individuals to assert their power and authority over others.

(***Source***: Andal, 1998).

To summarize, the communication theory proposes that every living entity needs to communicate with others and among themselves to survive.

Types of Theories

There are four main types of theories; these include: commonsense theories, working theories, scholarly theories and normative theories (Andal, 1998).

1. Commonsense Theories: Commonsense, as per Merriam Webster Dictionary, is practical judgment based on a simple understanding of facts and situations. It's the basic knowledge and judgment most people possess or believe they should have, acquired from personal experiences, and advice from others, and is crucial for safe and reasonable living. While commonsense ideas often align with human experiences, they can be formed from personal experiences or passed down wisdom from friends and family. These ideas guide our communication decisions but can occasionally lead to misunderstandings.

2. Working Theories: These are profession-specific generalizations about the most effective methods for accomplishing a task. For example, journalists follow the inverted pyramid structure, placing the most crucial information at the beginning. These theories serve as practical guidelines rather than formal, systematic models. They are commonly taught in specialized courses like public relations, media production, or public speaking. These working theories encompass established practices, traditions, and conventions that provide continuity to media production efforts.

3. Scholarly Theory: It is a common misconception that scholarly theories are irrelevant for business and professions, but the term "scholarly" indicates that these theories are based on rigorous research. These theories, also known as social scientific theories, are derived from systematic observations and objective analyses of mass media variables, using methods akin to empirical research in social sciences. As a result, scholarly theories offer more comprehensive, precise, and abstract explanations for communication compared to commonsense or working theories. However, they can be more intricate and challenging to grasp. Nevertheless, if you're genuinely dedicated to deepening your understanding of communication, scholarly theories provide a solid foundation for doing so.

4. Normative Theories: Normative theories of the press focus on the relationship between media and government. They seek to understand how the societal and political environment influences media structure and performance. These theories propose that the press mirrors its social and political context. Normative media theory encompasses the philosophical principles guiding the interaction between media and society. It concerns what the media should be doing for society, rather than what they currently do. The dominant ideas about mass media's responsibilities are typically in line with a society's core values. For instance, in the Western liberal tradition, this includes concepts like freedom, equality, social cohesion, cultural diversity, active participation, and social responsibility. While the landscape of media has evolved with new forms, normative theories remain relevant in shaping ideals about media's role in society and the extent of legitimate intervention to safeguard public interest. These theories establish norms and standards to assess media actions and set expectations for their structure, behavior, and performance.

The main relevant variants can be described as follows:

a. Authoritarian Theory: Media is controlled by ruling authority; dissenting opinions can be censored to maintain social order, even at the expense of free expression.

b. Free Press Theory: It Advocates unrestricted freedom of expression and economic operation of media, opposing government interference. Believes a functioning market can address media responsibilities and societal needs.

c. Social Responsibility Theory: It builds on free press theory, stressing media, particularly broadcasting, and accountability to society. While media remain free, they should also fulfil obligations for the public good. These responsibilities can be enforced through professional self-regulation, public intervention, or a combination of both.

d. Development Media Theory: It prioritizes economic, social, and political development over absolute media freedom in resource-constrained, economically developing countries.

e. Democratic Media Theory Favours small-scale, participatory, non-commercial media representing grassroots perspectives. Their role is to advocate for marginalized groups and maintain radical criticism.

f. Soviet Communist Media Theory: Media is state-controlled, serving as the voice of both the government and the people. Its purpose is to support and perpetuate the socialist revolution, defend against counter-revolution, and shield from capitalist influence.

Conclusion

From the above content and facts, it can be concluded that effective communication is a linchpin for personal and career success, emphasizing its crucial role in understanding, collaboration, and societal cohesion. In the digital era, communication plays the most important role in the significance of technological integration. Furthermore, it delineates various communication models, each offering distinct insights into the message transmission process. Additionally, the chapter distinguishes between different types of communication, such as verbal and non-verbal, formal and informal, as well as intra and interpersonal. This categorization provides a nuanced understanding of the multifaceted nature of human interaction. It introduces significant communication theories, offering valuable frameworks for comprehending how information is exchanged, interpreted, and utilized. These theories span from commonsense and working theories to scholarly and normative theories, each contributing to our understanding of communication dynamics in various contexts. Overall, this chapter underscores effective communication's profound impact on our personal and professional lives. It serves as a cornerstone for building meaningful connections, resolving conflicts, and advancing societal progress. As we navigate an increasingly interconnected world, the ability to communicate skilfully remains an indispensable skill for personal and career success.

References

Andal, N. (1998). Communication Theories and Models, Himalaya Publishing House. https://books.google.co.in/books?id=cD0lzQEACAAJ

Board Infinity. (2023). Formal vs. Informal Communication. Stable URL: https://www.boardinfinity.com/blog/formal-vs-informal-communication/#:~:text=Formal%20communication%20is%20also%20known,communication%2C%20telephonic%20conversations%2C%20etc.

ICAR e-Cource. (2012). Defination of Communication, Fundamentals of Extension Education.http://ecoursesonline.iasri.res.in/mod/page/view.php?id=15403#:~:text=Communication%20is%20the%20discriminatory%20response,stimuli%20(Hastman%2C%201966)

Leagans J.P. (1961). The Communication Process in Rural Development. 6(6), Comparative extension publications, Cornell University.

Suresh, K. (2003). Mass Communication Theories of Communication-PEOI Journalism and Mass Communication. https://www.peoi.org/Courses/Coursesen/mass/mass2.html

TNAU. (2012). Communication. Development of e-Courses for B.Sc (Agriculture). IASRI (Indian Agricultural Statistics Research Institute). http://ecoursesonline.iasri.res.in/Courses/Extension%20Methodologies%20for%20Transfer%20of%20ag.%20Tech/Data%20Files/lect01.html

TNAU. (2012). Defination of Communication. Development of e-Courses for B.Sc (Agriculture). IASRI (Indian Agricultural Statistics Research Institute). http://ecoursesonline.iasri.res.in/mod/page/view.php?id=15403#:~:text=Communication%20is%20the%20discriminatory%20response,stimuli%20(Hastman%2C%201966)

15

Agricultural Innovation through Market-Led Extension

Badigi Pavan Naik[1*], Bijaylaxmi Behera[2], Fakir Mohan Prusti[3] and Indira Priyadarshini Pattnaik[4]

[1,4]PhD Scholar, Department of Agriculture Extension Education, Odisha University of Agriculture and Technology, Bhubaneswar.
[2]Research Scholar, Department of Extension Education and Communication Management
Dr. Rajendra Prasad Central Agricultural University Pusa, Samastipur-848125, Bihar, India
[3]Research Scholar, Department of Agricultural Extension Education, Orissa University of Agriculture and technology, Bhubaneswar, Odisha.
[*]Corresponding author's e-mail: pavanextension@gmail.com

Abstract

Market-led extension is a dynamic approach to agricultural and rural development that places the needs and aspirations of farmers and rural communities at the forefront of its strategies. This concept recognizes that effective agricultural extension services must be responsive to market dynamics and the evolving demands of consumers, rather than relying solely on traditional knowledge transfer. Market-led extension involves facilitating the exchange of information, technology, and best practices between farmers and markets, enabling them to adapt to changing market conditions and consumer preferences. However, this approach also presents challenges, such as the need for strong public-private partnerships, adequate infrastructure, and access to finance for farmers. Additionally, ensuring inclusivity and equitable benefits for all farmers, especially smallholders, remains a critical challenge in the implementation of market-led extension programs. Overcoming these hurdles is essential for harnessing the full potential of this concept to drive sustainable agricultural development and improve rural livelihoods.

Keywords: *Agriculture, Challenges, Development, Extension Personnel, Farmers.*

Introduction

Market-led extension is a strategic approach aimed at providing farmers with valuable and high-quality information to enhance their decision-making processes in both production and marketing. This approach acknowledges the global nature of modern markets and encourages farmers to transition from being mere producers in local markets to becoming sellers in broader, potentially international markets. To make this shift successful, it is essential to equip extension workers, those who offer agricultural guidance, with the necessary skills and knowledge about market dynamics (Ahmad and Slathia, 2011). This involves a shift in their approach, moving from solely focusing on increasing productivity to emphasizing profitability. It also involves transitioning from subsistence farming to a more commercial orientation, from single-commodity farming to diversified crop production, and from local markets to global export markets. Beyond traditional agricultural concerns, market-led extension encompasses a broader range of topics, including product quality, consumer preferences, market trends, processing techniques, value addition, and other aspects of marketing. By empowering farmers with this information, the goal is to help them achieve higher returns for their agricultural products while simultaneously reducing production costs and enhancing the value and marketability of their goods. Crucially, market-led extension also prioritizes sustainability (Shitu et al., 2013). It encourages farmers to strike a balance between profitability and environmental and social responsibility to ensure that the needs of future generations are not compromised. In essence, market-led extension is a contemporary approach that recognizes the evolving nature of agricultural markets and seeks to empower farmers with the knowledge and skills needed to thrive in today's competitive and interconnected agricultural landscape.

Concept of Market-led Extension

In the evolving landscape of agriculture, it is imperative to equip farmers with timely and relevant information spanning from advanced production techniques to market-driven insights. This encompasses crucial questions such as what to cultivate, when to plant, how much to yield, when and where to market their products, at what price, and in what form to present their produce. Essentially, the concept of market-led extension regards farmers as agripreneurs, encouraging them to view farming as a business endeavour and enabling them to maximize monetary returns from their agricultural activities (Banerjee et al., 2022). Moreover, market-led extension offers farmers a diverse array of best practices tailored to suit their specific local conditions and farming systems. By doing so, it ensures that farmers have access to a versatile toolkit of strategies and approaches that align with their unique agricultural contexts, ultimately fostering success in their farming enterprises.

Objective of Market-led Extension

1. Identifying potential areas for agricultural marketing intervention by extension services.
2. Developing and implementing effective extension methods to offer tailored assistance to the farming community in marketing their agricultural products.
3. Expanding the existing Research-Extension-Farmers connection to include market connections.
4. Recognizing and disseminating innovative value-added practices and post-harvest technologies.
5. Establishing networks and formulating action plans within extension services to facilitate product marketing across various levels.

Prospects of Market-led Extension

The agricultural extension system must undergo a significant transformation. Its role needs to shift from merely transferring production technologies and ensuring the adoption of farming practices to actively promoting agriculture as a thriving agribusiness enterprise. Market-led extension holds substantial potential in enabling sustainable, high-yield agricultural production, especially in light of the current global challenges facing food production. For many years, there has been a strong emphasis on the transfer of knowledge from laboratories to actual farming practices, often referred to as 'lab to land.' However, it is now critical to shift our focus towards 'farm to fork.' This means that we need to consider the entire agricultural value chain, from cultivation to the ultimate consumer, to ensure efficiency, quality, and competitiveness. With the influence of international trade agreements like the World Trade Organization (WTO), countries worldwide are no longer limited to domestic production. Nations with a competitive edge are eager to export their agricultural output to global markets. In this changing landscape, the role of extension personnel in market-led extension becomes pivotal. Their ability to equip farmers with market-oriented knowledge and strategies can play a decisive role in the continued development and success of agriculture in our country (Joshi et al., 2012). This shift towards market-led extension has the potential to drive agricultural advancement and address the challenges associated with food production in a global context.

Role of extension personnel in light of Market-led Extension

1. Conducting a SWOT analysis of the market to raise awareness among farmers and aid in their production and marketing planning.
2. Forming Farmers' Interest Groups (FIGs) based on specific commodities and strengthening their capacities.

3. Improving farmers' interactive and communication skills to engage with customers and intermediaries (middlemen), facilitating feedback exchange and enhancing negotiation capabilities during direct marketing.
4. Establishing connections between farmers' groups, markets, and private processors for marketing and agro-processing purposes.

To effectively implement these strategies, it's essential to tailor them to the specific needs and conditions of the local farming community. Regular assessments and adjustments should be made to ensure that these initiatives are helping farmers improve their production and marketing capabilities. Additionally, collaboration with government agencies, NGOs, and agricultural experts can provide valuable support in implementing these strategies successfully.

Shifting of paradigm from production-driven to market-driven

1. Technology Transfer

Communication Approach: Persuasive and top-down.

Objective: To recommend specific farming practices and technologies to farmers and persuade them to adopt these recommendations.

Characteristics: In this paradigm, extension workers take the lead in providing recommendations and convincing farmers to adopt prescribed practices and technologies without substantial farmer involvement in decision-making.

2. Advisory Work

Communication Approach: Persuasive and participatory.

Objective: Farmers initiate communication by seeking advice or raising questions. Extension workers respond by offering guidance based on the farmers' expressed needs.

Characteristics: Advisory work is more interactive, with farmers actively seeking information and extension workers providing advice and recommendations in response to farmers' inquiries.

3. Human Resource Development

Communication Approach: Educational and mentorship-based.

Objective: To educate and empower farmers by enhancing their knowledge and skills, with extension workers taking on a mentorship role.

Characteristics: This paradigm centers on educating and improving farmers' capabilities. Extension workers serve as educators and mentors, offering training and support to enhance farmers' skills and knowledge.

4. Facilitation of Empowerment

Communication Approach: Empowering and collaborative.

Objective: To empower farmers by involving them in decision-making, problem-solving, and community development initiatives.

Characteristics: The empowerment paradigm emphasizes collaboration and partnership between extension workers and farmers. Extension workers facilitate processes where farmers actively participate in making decisions that affect their farming and broader community development.

These paradigms offer distinct approaches to agricultural extension, each with its unique communication style and underlying philosophy. The selection of a particular paradigm depends on the goals of the extension program, the desired level of farmer engagement, and the specific context and needs of the target farming community.

This paradigm has its roots in universities providing extension services to those unable to access higher education, offering knowledge and skills through education. It empowers students to make independent decisions based on the knowledge they acquire. However, this approach still maintains a top-down, paternalistic nature.

The latest paradigm, "facilitation of empowerment," is both educational and participatory. It's educational as it imparts new information, practices, and skills for enhancing production and environmental protection. It's participatory because it promotes communication among farmers themselves and encourages interactive processes to educate them. As agricultural extension evolves from one paradigm to another, the role of farmers and stakeholders evolves accordingly. In the "technology transfer" paradigm, development is achieved by disseminating modern research findings to traditional farmers, treating them as passive recipients of predetermined technology. In the "advisory works" paradigm, farmers' ability to identify and seek solutions to their problems is recognized, but the solutions provided are still based on pre-established packages, limiting farmers' choices. In "human resource development," extension workers determine what knowledge, skills, and technology to impart, but they also equip farmers to independently assess technical information and make responsible decisions. In the "facilitation for empowerment" paradigm, extension workers are no longer viewed as exclusive experts with all the answers (Mukherjee et al., 2022). Instead, farmers' knowledge, skills, and creativity are acknowledged as valuable resources, and problem-solving become a collaborative effort between extension workers and farmers. The shift in extension paradigms is primarily driven by the changing needs of rural communities aligned with the government's agricultural development goals. As one goal is accomplished, the government shifts its focus to the next, prompting corresponding changes in extension paradigms. Ultimately, national food security, a government's ultimate goal, is achieved and maintained through the transfer of research findings from researchers to farmers. There is a clear indication of the extension paradigm shifting towards "facilitation for empowerment." Current extension efforts focus on equipping farmers and involving them in planning and decision-making processes. In the current scenario, the extension's role expands beyond safeguarding national food security to promoting commercialization and export crop production. This involves empowering farmers with traditional skills, enhancing their understanding of local conditions, providing training in management and decision-making, fostering leadership and organizational skills, and improving marketing, negotiation, and technology utilization abilities.

Challenges in Market-led Extension

The public extension system is extensive and already deeply involved in numerous activities. Introducing a new marketing dimension into this system presents several hurdles. Firstly, it's essential to motivate and train extension personnel to acquire the requisite marketing skills before they can effectively engage in marketing extension activities.A significant challenge in market-led extension is the creation of accurate, relevant, comprehensive, and timely market intelligence. This intelligence encompasses various aspects, including product information, market practices, customer behavior, and commodity prices. The task of obtaining and maintaining up-to-date market intelligence can be quite formidable. To address these challenges, there is an urgent need to formulate an information policy aimed at ensuring that farmers have access to valuable information. This may involve revitalizing websites and digital platforms to provide farmers with accessible and pertinent data. Furthermore, the disparities in transport facilities, storage options, and warehousing capabilities across different regions pose additional noteworthy challenges in the implementation of market-led extension efforts. It is crucial to address these challenges comprehensively to ensure that all farmers have equitable access to markets and essential resources.

Conclusion

Market-led extension strives to empower farmers to realize greater returns for their agricultural produce while minimizing production costs, enhancing product value, and improving marketability. To fully leverage this approach, it's essential to tap into the potential of information communication technology and utilize various media channels, including electronic and print media, to disseminate crucial production and market-related information to farmers. In India, farmers have transitioned from subsistence farming to achieving self-sufficiency, largely due to the adoption of advanced production technologies. To succeed in the current liberalized market environment, the focus must shift from being 'supply-driven' to 'market-driven.' This means farmers should align their production with market demands, allowing them to earn higher returns on their agricultural investments.

By adjusting production according to market needs and staying well-informed about market dynamics, farmers can optimize their agricultural practices, boost profitability, and effectively navigate the complexities of today's agricultural markets.

References

Ahmad, N., and Slathia. (2011). Market Led Extension for Promoting Rural Entrepreneurship in India. Journal of Global Communication, 4(2): 143-147.

Banerjee, A., Rampal, V.K., & Ray, P. (2022). Knowledge Level of DAESI and Non DAESI Dealers for Paddy and Wheat Cultivation in Punjab, Indian Journal of Extension Education, 58(3): 42-45.

Joshi, P.V., Bhujbal, M.B., & Sarita B. (2012). Marketing Decision Model: A Market Led Extension technique for Efficient Agricultural Marketing. Indian Streams Research Journal, 2(3): 1-4.

Mukherjee, S., Kumari, P. H., Agrawal, S.S., Nagendran, S., & Suravajhala, R. (2022). Omics in medicinal plants, Elsevier BV.

Shitu, G. A., Sakia, A. R., Meti, S. K. and Maraddi, G. N. (2013). Market led extension; prospects and challenges for Agricultural sustainability in the 21st century. Conference Paper.